Keywords — Welding, health, cancer, disease, exposure, fumes, gases, literature review, noise, radiation, toxicology

Effects of Welding on Health — VII

Research performed by Biomedical Toxicology Associates, Rockville, Maryland, under contract with the American Welding Society and supported by industry contributions.

An updated (January 1986-December 1987) literature survey and evaluation of the data recorded since the publication of the first report (1979). This series of reports is intended to aid in the understanding of the health effects of welding.

Performed by:

Winifred Palmer

July 1989

Abstract

This literature review, with 194 citations, was prepared under contract to the American Welding Society for its Safety and Health Committee. The review deals with studies of the fumes, gases, radiation, and noise generated during various arc welding processes. Section 1 summarizes recent studies of the occupational exposures, while Section 2 contains information related to the human health effects of exposure to byproducts of welding operations. Section 3 discusses studies of the effects of welding emissions on laboratory animals and in vitro cell systems. Referenced materials are available from Biomedical Toxicology Associates.

Prepared for
Safety and Health Committee
American Welding Society
550 N. W. LeJeune Road, P.O. Box 351040
Miami, Florida 33135

International Standard Book Number: 0-87171-344-6

American Welding Society,
550 N.W. LeJeune Road, P.O. Box 351040, Miami, Florida, 33135

This report is published as a service and convenience to the welding industry and is the product of an independent contractor (Biomedical Associates) which is solely responsible for its contents. The materials in this report have not been independently reviewed or verified and are offered only as information. AWS assumes no responsibility for any claims that may arise from the use of this information. Users should make independent investigations to determine the applicability of this information for their purposes.

Personnel

AWS Safety and Health Committee

K. L. Brown, Chairman	The Lincoln Electric Company
K. A. Lyttle, 1st Vice Chairman	Union Carbide Corporation
R. J. Simonton, 2nd Vice Chairman	INCO Alloy International, Incorporated
M. E. Kennebeck, Jr., Secretary	American Welding Society
*O. J. Fisher**	Consultant
J. F. Hinrichs	A. O. Smith Corporation
D. G. Howden	Ohio State University
W. S. Howes	National Electrical Manufacturers Association
A. F. Manz	A. F. Manz Associates
M. T. Neu	Caterpillar Tractor Company
*G. R. Spies, Jr.**	Consultant
M. J. Vasquez	General Motors Corporation

* Advisor

Foreword

(This Foreword is not a part of *Effects of Welding on Health VII*, but is included for informational purposes only.)

This literature review was prepared for the Safety and Health Committee of the American Welding Society to provide an assessment of current information concerning the effects of welding on health, as well as to aid in the formulation and design of research projects in this area, as part of an ongoing program sponsored by the Committee. Previous work consists of the reports *Effects of Welding on Health* (I through VI) each covering approximately 18 months to two years. Conclusions based on this review and recommendations for further research are presented in the introductory portions of the report. Referenced materials are available from Biomedical Toxicology Associates.

Comparative Listing — Welding Processes

Explanatory Note: Terms used in the technical literature sometimes do not correspond to those recommended by AWS in its publication ANSI/AWS A3.0, *Standard Welding Terms and Definitions.*

Accordingly, the following list may aid the reader in identifying the process in use.

EWH — VII		Preferred AWS Term
Gas or Flame Cutting	(OC)	Oxygen Cutting or (OFC) Oxyfuel Gas Cutting
Gas Welding	(OFW)	Oxyfuel Gas Welding or (OAW) Oxyacetylene Welding
MAG	(GMAW)	— (with specified shielding gas)
MIG, GMA	(GMAW)	Gas Metal Arc Welding
MMA, SMA	(SMAW)	Shielded Metal Arc Welding
TIG	(GTAW)	Gas Tungsten Arc Welding
Wire		Electrode

Acknowledgments

Funds for this project were provided by the American Welding Society.
The American Welding Society gratefully acknowledges the financial support of the program by industry contributions.

Supporting Organizations

Air Products and Chemicals, Incorporated
Airco Welding Products
Allis-Chalmers
Alloy Rods Division, The Chemetron Corporation
AWS Detroit Section
AWS New Orleans Section
Arcos Corporation
The Binkley Company
Caterpillar Tractor Company
Chicago Bridge and Iron Company
Grove Manufacturing Company, Division of Kidde, Incorporated
General Electric Company
The Heil Company
Hobart Brothers Company
INCO Alloys International
Lincoln Electric Company
Miller Electric Manufacturing Company
National-Standard Company
A.O. Smith Corporation
Teledyne-McKay, Incorporated
Trinity Industries, Incorporated
Truck Trailer Manufacturers Association
Walker Stainless Equipment Company
Weld Tooling Corporation

Many other organizations have also made contributions to support the ongoing program from May 1979 to the present.

Table of Contents

List of Tables

List of Figures

Introduction

The health of workers in the welding environment is a major concern of the American Welding Society. To stay abreast of this subject, the health literature is periodically reviewed and published in the report Effects of Welding on Health. Six volumes have been published to date; the first covered data published before 1978, while the latter five covered time periods between 1978 and December 1985. The current report includes information published between January 1986 and December 1987. It should be read in conjunction with the previous volumes for a comprehensive treatment of the literature on the Effects of Welding on Health.

Included in this volume are studies of the characteristics of welding emissions that may have an impact on control technologies necessary to protect the welder (Section 1). In keeping with previous volumes, the health studies are organized according to the affected organ system. The respiratory tract, the primary route of exposure to welding emissions, is also a major target organ of many components of these emissions. Acute (e.g., metal fume fever, cadmium poisoning) as well as potential chronic respiratory effects (e.g.,emphysema, cancer) of welding emissions are of concern. However, chronic effects are not as well defined or understood, and the risk of cancer from these exposures has not been clearly established. Continued research in the form of epidemiologic studies, investigations with laboratory animals, and in vitro genotoxicity studies will help to resolve this question.

Executive Summary

The Respiratory Tract

Much of the recent research performed on the health effects of welding involved the respiratory tract. Of major concern was whether exposure to welding emissions induces conditions such as reduction in pulmonary function, bronchitis, and emphysema. Essential to this question is whether pulmonary impairments are associated with exposure to specific component(s) of welding emissions and the identification of welding electrodes that produce less hazardous emissions.

Rounded opacities or shadows are frequently observed in chest X-rays of welders. These opacities represent deposits of fume particles in the respiratory tract and are not usually associated with loss of pulmonary function. Larger and coalescent shadows reflect fibrotic processes and do not usually result from welding exposures. X-ray shadows often regress after welding exposure ceases. A recent study by Desmeules and Tardif indicated that the continued progression of X-ray opacities after welding exposures ceased is associated with exposure to lung irritants from sources other than welding (Ref. 33).

Some investigators have reported that a mild reduction in pulmonary function accompanies the appearance of nodular shadows in X-rays. However, there are reports to the contrary and other investigators consider pneumoconiosis in welders to be a benign condition. The literature shows many disparities among the results of lung function tests in welders. During this report period, deficits in pulmonary function among welders were reported in two studies, while two others found no correlation between welding exposures and abnormalities in pulmonary function. These disparities may result from the use of tests that measure the function of those parts of the respiratory tract least likely to be affected by welding fumes. Other disparities arise from studies of heterogeneous populations. The quantities and chemical characteristics of gaseous and particulate components of welding emissions vary widely with the welding method. Since the effects on the respiratory tract can be expected to vary as well, it is important to correlate observed effects on the respiratory tract with the welding method used.

Cancer

Thirteen new cancer epidemiology studies were conducted which included welders among the study populations. In most of these, the inclusion of welders was incidental and little information could be obtained about the duration of exposure, type of welding process used, or possible important nonwelding exposures. Only one study specifically examined a cohort of welders; no excess cancer of any type was found among the 3000 welders examined (Ref. 25). Of the eight studies which evaluated lung cancer rates, four found a significant association between welding and lung cancer. In three of these, exposures to carcinogens from nonwelding processes could not be ruled out (Refs. 90, 152, 158, 161, and 162).

An elevated risk for leukemia was noted in a shipyard study performed by F. B. Stern (Ref. 169). In contrast, R. M. Stern found no association between welding exposures and leukemia in his analysis of 15 epidemiologic studies which considered the incidence of leukemia in welders (Ref. 170). Similarly, an elevated risk for brain tumors was found by McLaughlin in a population-based study in Sweden (Ref. 97), but no brain tumors were identified in a death- certificate study by Thomas in the U.S. (Ref. 175), or in a case-control study conducted by Olin in Sweden (Ref. 123). Isolated studies reported elevated risks for cancer of the kidney (Refs. 158 and 161), gastrointestinal tract (Ref. 141), nasal sinus (Ref. 52), and pancreas (Ref. 109).

A major argument supporting the probability of an increased cancer risk in some welders is the presence of nickel and hexavalent chromium in fumes from some welding processes. The mechanism of action of metal carcinogens remains unknown and is an important area for further research. Epidemiologic studies differ widely in their conclusions concerning the cancer risk of welders. Most of the 13 cancer epidemiology studies reviewed in this volume of Effects of Welding on Health were of general populations, making it difficult to rule out nonwelding exposures. More carefully controlled studies of welding populations are critically needed to help understand the risk of cancer in welders. Studies of large populations of welders, in which exposures to specific welding processes can be carefully delineated are necessary to provide a much needed answer to the question of whether exposures to specific fumes account for the elevated cancer risk observed among some welding populations.

Effects on the Urogenital Tract

With the exception of cadmium-induced kidney damage, kidney disease is not generally associated with welding. This was confirmed by two epidemiologic studies which found no relationship between welding and the incidence of chronic kidney disease (Refs. 51 and 193). However, chronic exposure to low concentrations of cadmium fumes can cause kidney damage, which is first manifested by urinary excretion of low molecular weight serum proteins. Signs of kidney tubular dysfunction were found among workers exposed to cadmium in welding and soldering fumes and aerosols from cadmium-containing paints (Refs. 164 and 184). A study of solderers with a work history of high cadmium exposures and signs of kidney tubular dysfunction suggested that mild tubular proteinuria will not progress to more severe kidney dysfunction in the absence of further exposure to cadmium (Ref. 164).

Biological Monitoring

Because hexavalent chromium passes more readily through cell membranes, trivalent chromium tends to be transported by plasma while hexavalent chromium is transported mainly in erythrocytes (Ref. 3). Based on the possibility that this difference could be used to distinguish between exposures to trivalent and hexavalent chromium, two research groups examined chromium levels in plasma and red blood cells of stainless steel welders. Angerer et al. (Ref. 4) concluded that simultaneous determination of chromium levels in plasma and erythrocytes may give a useful picture of overall exposure to both total chromium and hexavalent chromium. In contrast, Gustavsson and Welinder (Ref. 50) concluded that determination of blood or erythrocyte chromium levels would not add important information to that gained from biological monitoring of urinary chromium concentrations. Further studies are needed to establish the utility of chromium distribution between plasma and red blood cells in biological monitoring.

Stability of Welding Fumes

Recent work revealed that particles in some welding fumes undergo chemical and morphological changes in the lung. Physicochemical studies by Stern et al. (Ref. 172) indicated that substituted magnetite is the primary magnetic component in freshly formed shielded metal arc welding (SMAW) fumes whereas a less substituted form of magnetite appeared to be the predominant magnetic component of fumes that have been retained in the lungs. Similar findings were reported by Kalliomaki et al. (Ref. 76). They conducted an in vivo rat study which showed that the remanent magnetic fields of fumes generated by SMAW of stainless steel and gas metal arc welding (GMAW) of mild steel are not stable in the lungs.

Other in vivo studies showed that there is a marked difference in pulmonary clearance rates of fumes from SMAW and GMAW of stainless steel (Refs. 70, 71, and 73). Anttila demonstrated that these differences are related to differences in the solubility of particles in the lung (Refs. 6 and 7). The morphology and chemical composition of particles were examined periodically in rat lungs after exposure to welding fumes by inhalation. The majority of particles in SMAW fumes had irregular shapes and were enriched with sodium and potassium. These particles changed with time, losing practically all sodium, potassium, and chromium, and also some manganese. Eventually, so much material was lost that the particles became transparent. In contrast, the particles from GMAW fumes showed no signs of dissolution. Thus, the solubility of particle populations may be an important determinant in the pulmonary clearance of welding fumes.

Technical Summary

The Exposure

Fumes

A major objective of research performed on the composition of emissions from different electrodes and welding processes is to identify those methods which generate the least harmful pollutants. Such information is also useful for selection of electrodes for use in the workplace. Many studies of the composition of fumes, and the rate of fume emission from different welding processes, were published in 1986 and 1987.

An industrial hygiene survey conducted in six Dutch factories showed that the fume and gas exposures from plasma welding of stainless steel were comparable to the low exposures produced by gas tungsten arc welding (GTAW). Exposures from plasma cutting of stainless steel were comparable to those produced by SMAW (Ref. 183). Henderson et al. showed that the fume generation rate (FGR) was lower for gas shielded flux cored electrodes than for those used without gas shields (Ref. 58). With resistance welding, the quantity of emissions increased markedly during "splashing" and also when welding greased surfaces (Ref. 85). Studies of the effects of hyperbaric pressure on fume emission demonstrated that the ratio of manganese to iron increases, and the FGR is 5 to 10 times greater when performing GTAW at a pressure of 32 bar than at 1 bar (Ref. 16).

McIlwain and Neumeier determined FGRs and fume constituents generated by selected coated electrodes commonly used in mines (Ref. 96). Olah and coworkers quantified gases (ozone, hydrogen fluoride, and nitrogen oxides) and light and heavy elements [Fe, Cr, Cr(VI), Ni, Mn, Mo, Si, C, and others] in fumes produced by coated electrodes, flux cored hardfacing electrodes (Refs. 115, 116, 118, and 119), and by electrodes used for welding of high alloy steels (Refs. 117 and 177). They found that exposures resulting from hardfacing are comparable to those from SMAW of high alloy steels (Ref. 116). Tandon et al. determined 19 elements in the flux and fume from three types of hardfacing and two types of high strength, low alloy steel coated electrodes. An inverse relationship between the fluoride concentration in the fume and the FGR was found. There was a positive relationship between sodium and potassium concentrations in the flux and the ratio of water- soluble hexavalent chromium to total chromium in the fume (Ref. 174).

The concentration of hexavalent chromium in GMAW fumes is substantially greater in fume samples collected by wet impingement than in those collected by traditional filtration methods (Ref. 137). A study by Hewitt and Madden of GMAW fumes collected by wet impingement suggested that reactions between trivalent chromium oxides and ozone may be responsible for the formation of hexavalent chromium (Ref. 59). Wiseman and Chapman (Ref. 190) found traces (0.1 ppb) of nickel carbonyl in fumes produced by GTAW of nickel alloys with shield gases containing carbon monoxide or carbon dioxide (CO_2).

Several research groups described methods for determination of chromium and other elements in welding fumes by X-ray fluorescence analysis (Ref. 106, 120, and 186). Sawatari and Serita (Ref. 148) designed a method using a series of leaching solutions to differentiate between chromium of different valences and solubilities. Methods employing high pressure liquid chromatography (Ref. 94) and atomic absorption spectrometry (Ref. 21) to analyze elements in welding fumes were described by other investigators. Three research groups published methods for measuring trace quantities of metals in biopsied lung tissue (Refs. 7, 10, 88, and 89).

Particles

Welding fumes are composed of microscopic particles which tend to be of respirable size. The outer surface of particles from SMAW fumes are composed primarily of lighter elements derived from the flux (e.g., Na, K, Ca, F), while heavier elements derived from the filler metal (e.g., Fe, Cr, Ni, and Mn) tend to predominate in the particle core (Refs. 48 and 187). Kalliomaki et al. (Refs. 47, 49, and 72) classified particles in fumes from SMAW, GMAW, and flux cored arc welding (FCAW) of mild steel into four groups on the basis of their morphology: clusters of globules, chains, granular networks, and large spheres. The distribution of the particle classes varied with the welding process.

Ahlberg et al. determined the mass median aerodynamic diameter (MMAD), the equivalent mobility diameter (DME), and the growth in particle size as a function of relative humidity for particles in fumes from SMAW of mild steel (Ref. 2). They concluded that hydrated particles could collapse to spheres after entering the respiratory tract which would cause the MMAD to increase and the DME to decrease. This would result in increased sedimentation and diffusion rates, and would thereby cause greater particle deposition in the alveoli.

Gases

Engstrom and Virtamo determined ozone levels in the breathing zone of welders from 24 welding shops using different gas shielded welding processes (Ref. 39). The mean concentrations for all welding processes was less than the Finnish standard of 0.1 ppm. GMAW of aluminum produced the highest ozone concentrations with a geometric mean of 0.06 ppm. The addition of nitric oxide to the shield gas was less effective in reducing ozone levels than predicted by the laboratory studies of Sipek and Smars (Refs. 132 and 157).

Electromagnetic Radiation

Okuno estimated permissible exposure times at 1 meter from the source for the ultraviolet (UV) and blue light generated by various welding processes (Refs. 113 and 114). The permissible exposure times for UV light varied from several seconds to one minute for most of the welding processes studied. For each welding process, the intensity of the UV, but not the blue-light radiation, increased with the diameter of the wire, arc current, and voltage. As expected, the UV hazard was greater with GMAW than with SMAW. The greatest blue-light hazard was found for CO_2-shielded GMAW of mild steel, for which a permissible daily exposure time of 42 seconds was estimated. In similar work, Salsi and Barlier measured the spectral emissions from UV to far infrared light generated by 47 diverse coated electrodes (Ref. 9).

Production Coatings

Heikila et al. found that concentrations of benzo (a)pyrene as high as 0.64 micrograms/m^3 were present in the vicinity of welders during SMAW of railroad ties (Ref. 53).

Noise

Impulse noise may be more harmful to hearing than continuous steady state noise of the same energy level (Ref. 133). The impulse noise level has been defined as the difference between the peak value and the root mean square of the slow time-weighted value. Noise is considered to be impulsive if the difference between peak levels and the slow time-weighted value is greater than 15 dB. A study conducted in a shipyard assembly hall indicated that the impulsiveness for different processes increases in the following order: gouging $<$ carbon arc gouging $<$ chiseling $<$ welding (process not specified) $<$ GMAW (Ref. 168). Further tests indicated that earmuffs effectively attenuate impulsiveness (Ref. 138).

Hygiene and Work Practices

Accidents, the primary health risk associated with the use of robots, occur most often during teaching, testing, and maintenance. Accidents may result from pushing the wrong button in the robot controls, by improper programming of robot movement, or by difficulties in perceiving or anticipating the movement of the robot arm during work operations (Refs. 56, 57, and 134).

A fatal accident occurred in which a welder was working inside a vapor degreaser pit filled with 1,1,1-trichloroethane vapors. While welding, the vapors ignited and the welder fell into the pit, where he died. Tests of the 1,1,1-trichloroethane remaining after the accident indicated that it contained twice the quantity of corrosion inhibitors normally present, which may have contributed to the flammability of the vapors (Ref. 30).

Protective Gear. A study by Sliney et al. (Refs. 12 and 163) indicated that most fabrics used in standard work clothing provide adequate attenuation of the UV radiation generated by welding. Laundering decreased the UV transmission through most fabrics tested. This was thought to be due to shrinkage of the fabric, with a consequent decrease in the size of the openings between fibers.

A study by Beckett et al. (Ref. 11) demonstrated that wearing vapor-barrier garments in the workplace can cause heat stress. The authors noted that when heavy muscular work is performed in a warm environment, impairment of the dissipation of sweat by vapor-barrier garments can result in significantly higher body temperatures with a high risk for heat-related disorders, including heat stroke.

Jakubcik (Ref. 66) described a lightweight, maintenance-free, disposable respirator used in a steel works in Czechoslovakia to provide protection when welding high alloy steels with electrodes containing chromium, nickel, and manganese. Several methods for extracting welding fumes from the work area (Refs 67 and 145) and from hyperbaric chambers (Ref. 35) were published.

Training. Rosen and Lundstrom coupled video taping of workers and direct reading instrumentation of breathing zone exposures (Ref. 146). The resultant video tape of work practices with a superimposed presentation of real-time exposure data was used to evaluate the relationship between exposures to fumes, dusts or solvents with individual work patterns, and to train persons who work with hazardous substances.

Effects of Welding on Human Health

Respiratory Tract

Alveolar Macrophages. Pierre et al. (Ref. 140) examined the interactions of welding fumes with two lung antiproteinases. They found that fume particles bind reversibly to the antiproteinases and prevent them from inhibiting the activity of proteolytic enzymes.

Mylius and Gullvag showed that sputum from coke plant workers, aluminum plant potroom workers, asphalt workers, and shipyard welders contain more alveolar macrophages than controls. Combined smoking and occupational exposures produced a marked increase in the number of macrophages (Ref. 104).

Estimation of Retained Particles in the Lungs. Magnetopneumography utilizes the magnetic properties of inhaled particles to estimate the burden of occupational dusts in the respiratory tract. Kalliomaki et al. developed a magnetopneumograph that measures pulsed coercive force which can discriminate between remanent fields produced by different types of fumes (Ref. 75). An innovation introduced by Drenck and Stern (Ref. 36) used an AC susceptibility bridge to provide information about the total content of ferromagnetic, paramagnetic, and diamagnetic material. Using this system, Stern et al. (Ref. 172) obtained evidence for in vivo changes of welding fume particles from the substituted magnetite present in freshly formed welding fumes to unsubstituted magnetite. These findings were supported by studies with X-ray diffraction and Mossbauer spectroscopy which indicated that substituted magnetite is the primary magnetic component in freshly formed welding fumes, whereas a less substituted form of magnetite is the predominant magnetic component in inorganic dust obtained from ashed lungs of deceased welders.

A study in which fumes were administered to rats by intratracheal instillation also indicated that the magnetic properties of some welding particles are not stable in the lungs (Ref. 76). The specific magnetic moment of welding fumes generated by SMAW of mild steel and GMAW of stainless steel were stable, while the magnetic moment of welding fumes generated by SMAW of stainless steel and GMAW of mild steel decreased by about 50% during the 106-day period following the introduction of fumes into the lungs.

Pulmonary Function and Bronchitis. Chest X-ray abnormalities, and deficits in lung function, excess chronic bronchitis, or both, among welders were examined in 13 studies. Of these, two were confounded by exposure to asbestos or other nonwelding exposures in the workplace (Refs. 34 and 122).

Nodular opacities are frequently observed in chest X-rays of welders. These opacities are not usually associated with loss of lung function or diseases of the respiratory tract. During the current report period, two studies indicated that siderosis or pneumoconiosis in welders is not associated with a reduction of lung function (Refs. 26 and 166) while two other studies (Refs. 84, 150, and 151) concluded that such an association exists. Chest X-ray shadows often regress after welders retire or are removed from further exposure. Desmeules et al. found that workplace exposures other than welding existed in cases where opacities increased or progressed after welding exposures ceased (Ref. 33).

While bronchitis and loss of pulmonary function were attributed to welding exposures by several investigators, the reports differed in their conclusions concerning the contribution of tobacco smoke to the development of respiratory impairments. Five studies found that nonsmoking welders have an excess risk of bronchitis (Refs. 81, 82, 84, 92, 102, and 149) while two found an elevated incidence of bronchitis only among welders who smoked (Refs. 26 and 166).

Similar disparities exist among reports of the influence of tobacco smoke on the induction of impairments in lung function. Two studies found deficits in pulmonary function only in smoking welders (Refs. 26 and 166) while three other studies found an elevated incidence of pulmonary deficits in both smoking and nonsmoking welders. In a small cohort of shipyard welders, Kalliomaki found deficits in pulmonary function only among retired shipyard welders and not among the younger, active welders (Ref. 74).

Although there are conflicts among these reports, the results suggest that nonsmoking welders may develop bronchitis and deficits in pulmonary function. Welders who smoke clearly have an elevated risk for both bronchitis and pulmonary deficits. Whether the combined effects of smoking and welding exposures are additive or synergistic is not established. The study of shipyard welders by Lyngenbo et al. (Ref. 92) carefully controlled for exposure to cigarette smoke, asbestos and exposure to other lung toxicants. This study demonstrated that welding alone can cause bronchitis, upper respiratory tract symptoms, and reduced lung function. The effects of cigarette smoke on the development of these symptoms were additive.

The influence of constituents of welding fumes on the development of pulmonary impairments was examined by two research groups. Krasnyuk et al. found that welders of high alloy steel were more likely to have extensive pneumoconiosis with more severe restrictive disturbances in pulmonary function than welders of low alloy steel (Ref. 84). Ulrich et al. found no differences in pulmonary function between welders with high and low chromium exposures (Ref. 179).

Acute Respiratory Tract Disease — Case Report. The case of a welder who developed a sudden, work-related onset of pulmonary disease was described by Silberschmid (Ref. 156). Examination revealed swollen and red bronchial mucosa, irregular opacities in chest X-rays, and deficits in pulmonary function. Extreme inflammatory changes were found in biopsied lung tissue. Symptoms of exertional dyspnea, bronchial hyperreactivity, and recurrent pneumonias persisted for 7 years. Prior to the onset of disease, he had been working without properly functioning local ventilation and was exposed to high concentrations of fume.

Production Coatings. Emmerling et al. (Ref. 38) found no significant abnormalities in pulmonary function among 35 workers who welded metals primed with zinc chromate, polyvinylbutyral coatings or alkyl paints. In addition, there was no correlation between work place exposure and concentrations of zinc, hippuric acid, methylhippuric acid, formic acid, or carboxyhemoglobin in blood or urine samples. The number of participants in the study was small, and measurements of airborne chemicals indicated that exposures were low.

Cancer

Thirteen new cancer epidemiology studies were conducted which included welders among the study populations. Eight of these were case-control studies, while the remainder were population-based morbidity and mortality studies. Only one study specifically examined a cohort of welders. In that study, no excess cancer of any type was found among the cohort of 3000 welders (Ref. 25). Of the eight studies which evaluated lung cancer rates, four found a significant association between welding and lung cancer. In one of these, there was also a high incidence of mesothelioma and, thus, asbestos exposure could not be ruled out (Refs. 158, 161, and 162). A second study found an excess of lung cancer among shipyard welders, but not among welders who did not work in shipyards (Ref. 152), which suggests the effects of exposures other than welding. The remaining two were case-control studies in which the total number of welders with lung cancer were 28 (Ref. 83) and 19 (Ref. 90). In the latter, Lerchen cautioned that "the welders in our study had worked in diverse industries that may have involved exposures other than welding that also increased the lung cancer risk" (Ref. 90). Because of the possibilities of confounding exposures, it is difficult to draw firm conclusions concerning the lung cancer risk of welders on the basis of studies described above.

Other tumors reported to be associated with

welding include leukemia, cancer of the stomach, brain, nasal sinus, and pancreas. An elevated risk for leukemia was noted in a shipyard study performed by F. B. Stern (Ref. 169). In contrast, R. M. Stern found no association between welding exposures and leukemia in his analysis of 15 epidemiologic studies which considered the incidence of leukemia among welders (Ref. 170). Similarly, an elevated risk for brain tumors was found by McLaughlin in a population-based study in Sweden (Ref. 97), but no brain tumors were identified in a death-certificate study by Thomas in the U. S. (Ref. 175), or in a case-control study conducted by Olin in Sweden (Ref. 123). One study found an elevated risk for renal tumors in welders (Ref. 161), while isolated studies also reported elevated risks for cancer of the gastrointestinal tract (Ref. 141), nasal sinus, (Ref. 52) and kidney (Ref. 158). Norrell conducted a case control and morbidity and mortality study of pancreatic cancer (Ref. 109). An elevated incidence of pancreatic cancer in welders was observed in the case-control but not in the mortality study.

Screening for Cancer. A nontraumatic method for screening workers at risk of nasal cancer was reported by Reith et al. This cytologic screening method examines a representative population of cells obtained by brushing the surface of the middle nasal turbinate. In a study of nickel refinery workers with known nasal dysplasia, more cases of preneoplastic nasal lesions were detected by cytologic screening than by biopsy (Ref. 144).

Metal Fume Fever

Vogelmeier et al. (Ref. 185) observed changes in pulmonary function and marked pulmonary inflammation in a welder who had recurring symptoms of metal fume fever after welding zinc-coated materials, but not other metals. Abnormalities in pulmonary function and a marked increase in polymorphonuclear leukocytes in bronchoalveolar lavage fluid were associated with symptoms of metal fume fever.

A case of metal fume fever with a severe, concomitant allergic reaction to zinc fumes was reported by Farrell (Ref. 40). After welding galvanized metal, the patient experienced classic symptoms of metal fume fever associated with an immediate anaphylactoid reaction to zinc fumes as well as a delayed reaction consisting of angioedema and urticaria.

Effects on the Eye and Vision

Keratoconjunctivitis, a painful inflammation of the cornea, is the eye condition most frequently encountered by welders. Although this condition usually lasts for up to 2 days, Sharir and Ben-David observed severe symptoms of keratoconjunctivitis which persisted for several months in four workers employed in plasma arc thermal spray coating (Ref. 153).

Two cases of photic maculopathy, in which visual loss was noted soon after exposure to the welding arc, were reported (Refs. 20 and 180). In one of the cases, some visual loss and signs of retinal injury were present 6 months after the incident (Ref. 180).

An unusual case of retinal damage in a man with a lens implant was reported by Turut et al. (Ref. 178). Four years after the surgery, he welded without wearing eye protection and experienced an immediate loss of visual acuity. A small lesion, apparently caused by radiation from the arc, was present on the retinal fovea of the eye with the lens implant. There were no indications of damage or loss of vision in the eye with the normal lens.

Effects on the Skin

A case of a welder with recurrent, severe facial dermatitis was described by Shehade et al. (Ref. 154). Allergic dermatitis was ruled out by skin patch testing with a battery of allergens. The welder had routinely performed spot welding without a protective face mask. His case was diagnosed as photodermatitis, caused by UV exposure during welding.

Sensitivity to Fume Components

A case of a welder who had anaphylactoid reactions following exposure to chromates was reported by Moller et al. (Ref. 101). The symptoms included urticaria, difficulty breathing and chest tightness. Nordman et al. examined 58 stainless steel welders who had asthmatic reactions when welding stainless steel. Some of these welders responded positively to the histamine provocation test prior to welding and had significant reductions in pulmonary function following welding of stainless steel (Ref. 108).

Effects on the Musculoskeletal System

Two techniques for analyzing causes of postural stress were described. One was a computer-aided system used to analyze working postures of the trunk and shoulders on repetitive assembly line jobs (Ref. 78). Both systems were applied to improving postural stress incurred during welding operations (Refs. 55 and 78).

Effects on the Urogenital Tract

Chronic exposure to low concentrations of cadmium fumes can cause kidney damage, which is first manifested by urinary excretion of low molecular weight serum proteins such as beta-2-microglobulin and retinol-binding protein. Signs of tubular dysfunction, but not glomerular dysfunction, were found among a group of 26 cadmium-exposed workers by Verschoor et al. (Ref. 184). Smith et al. found low molecular weight proteins (typical of tubular dysfunction), but not albumin (characteristic of glomerular dysfunction) in the urine from 20 of 51 solderers with up to 25 years of uncontrolled exposure to fumes from cadmium-containing alloys. No significant changes in beta-2-microglobulin were found in 19 solderers followed for 2 years after cadmium exposure ceased, suggesting that mild tubular proteinuria will not progress to more severe kidney dysfunction in the absence of further exposure to cadmium (Ref. 164).

Two epidemiologic studies showed no association between renal impairment and welding. In the first, elevated levels of nickel and chromium were found in the urine of stainless steel welders, but kidney function, assessed by measurement of total protein and enzyme activity in urine, was comparable to that of the control group (Ref. 193). The second study examined kidney function in 143 males who had received kidney biopsies as a result of suspected or clinically established glomerulonephritis or obscure renal disease. There was no relationship between welding and the incidence of severe chronic kidney disease nor was there an increase in the risk of death from nephritis among welders (Ref. 51).

Effects on the Teeth

To investigate the cause of a metallic taste sensation and loosening dental amalgams among underwater welders, Ortendal et al. investigated the intraoral electrical and magnetic activity during underwater welding. Humidity was the predominant factor in the generation of intraoral electrical activity. However, even when water was deliberately introduced into the diver's suit, the intra- oral electric field was too small to damage dental amalgam. No correlation was found between the magnetic field and a metallic taste in the mouth. In vitro studies indicated that a slight degeneration of dental amalgams may be caused by magnetic fields equivalent to those generated by welding. However, chemical changes were insufficient to account for a metallic taste sensation.

Biological Monitoring

Sjogren et al. (Refs. 159 and 160) examined the relationship between urinary aluminum concentrations and exposure to fumes generated by GMAW and GTAW of aluminum. Urinary aluminum levels were dependent on the most recent daily exposures as well as on the total duration of exposure. The results indicated that aluminum is probably stored in two or more compartments in the body, with widely different clearance rates.

Two studies were based on findings that red blood cells are permeable to hexavalent, but not trivalent, chromium (Refs. 4 and 50). Angerer et al. (Ref. 4) determined chromium concentrations in the breathing zone and in erythrocytes, plasma and urine from 103 male welders who performed SMAW or GMAW, or both, of stainless steel. Chromium levels in erythrocytes were too low to provide accurate estimates of exposure below airborne concentrations of 100 micrograms/m^3. However, the investigators concluded that simultaneous determination of chromium levels in plasma and erythrocytes may give a useful picture of exposure to total and hexavalent chromium.

Different conclusions were drawn by Gustavsson and Welinder (Ref. 50) who observed no obvious differences in clearance rates of erythrocyte and plasma chromium in studies performed with four stainless steel SMA welders and six electroplaters. These investigators concluded that determination of blood or erythrocyte chromium levels would not add important information to that gained from biological monitoring with urinary chromium concentrations.

Zschiesche et al. (Ref. 194) determined levels of manganese in the breathing zone and body fluids of eight welders from two factories who worked with filler materials containing about 0.8% manganese. Levels of manganese in blood and urine tended to be low and, for the most part, did not vary substantially from values obtained for the general population. There were no correlations between external exposures and internal manganese levels, nor with the time of day or the day of the week that the biological samples were collected.

Effects on the Voice

A study by Ohlsson indicated that voice and throat problems were much more frequent among welders than office workers. The investigators concluded that vocal problems among welders are due to high noise levels which force a strain on the larynx during vocalization, whereas welders believed their problems resulted from fume exposure (Ref. 112).

Toxicologic Investigations in Animals and Cell Cultures

Animal Studies

Inflammation and Fibrosis. The development of practical methods for assessing the fibrogenicity of fumes and gases from welding operations is important for the determination of risk associated with different electrodes and welding methods. Hicks et al. (Ref. 62) compared the irritant and fibrogenic properties of fumes from different electrodes by injecting fume samples into rat and guinea pig muscle tissue. The inflammation produced by welding fumes decreased in the following order: SMAW of stainless steel >SMAW of mild steel >GMAW of stainless steel.

Studies of the effects of fumes on the respiratory tract were conducted by the American Welding Society (Ref. 23). Rats were exposed by inhalation for 6 hours to fumes from six welding processes and the histopathologic effects on the respiratory tract were determined. Other assays examined viability and phagocytic activity of alveolar macrophages from exposed animals. Extracts from urine were not genotoxic in the Salmonella/Ames test nor were any chromosome aberrations seen in bone marrow cells recovered from animals 20 hours after exposure. On the basis of these studies, the electrodes were classified into three categories: highly toxic (E308-16), moderately toxic (E7018, E6010, and E70T-1) and nontoxic (E70S-3 and E5356).

Wilmer et al. conducted 14-day inhalation studies of fumes from ten electrodes in Wistar rats (Ref. 188). Fumes from SMAW of stainless steel were the most toxic, causing severe dyspnea, some mortality, and reduction in body weight gain. GMAW fumes were less toxic and induced only slight signs of respiratory distress. Fumes from SMAW of mild steel were only slightly toxic. The fumes from most electrodes were cytotoxic to alveolar macrophages. None of the fumes caused sister chromatid exchanges (SCE) or structural chromosome aberrations in bone marrow cells.

Clearance. In vivo studies conducted by Kalliomaki et al. demonstrated marked differences in pulmonary clearance rates of fumes from SMAW and GMAW of stainless steel (Refs. 70, 71, and 73). Anttila showed that this was due to differences in the solubility of particles in the lung (Refs. 5 and 7). The morphology and chemical composition of particles were examined periodically after exposure of rats to welding fumes by inhalation. The majority of particles in SMAW fumes had irregular shapes and were enriched with sodium and potassium. These particles changed with time, losing practically all sodium, potassium, chromium, and some manganese. In contrast, the particles in GMAW fumes showed no signs of dissolution.

Another study focused on clearance of manganese from the respiratory tract (Ref. 77). The results suggested that manganese was cleared from the lungs as intact particles and that little manganese clearance resulted from in situ dissolution of particles.

Toxic Effects. A long-term inhalation study of the effects of nickel oxide and arsenic trioxide in Wistar rats was conducted by Glaser et al. (Ref. 42). Arsenate exposure had no apparent effects. Rats treated with nickel oxide had an increased mortality rate and severe alveolar proteinosis. No treatment-related tumors were seen.

Gorban et al. examined the effects of exposure to fumes generated by CO_2-shielded GMAW of medium and high alloy steel with two types of austenitic electrodes (Ref. 45). The LD50s did not differ significantly for fumes from the two electrodes, but differences were noted in the effects on levels of nucleic acid and alanine and aspartate aminotransferase in the liver and lung.

Geleskul et al. (Ref. 41) used the formation of malondialdehyde in the presence of ascorbic acid as a measure of lipid peroxidation produced by welding fumes. The investigators concluded that lipid peroxidation can be used for the assessment of the toxicity of welding fumes.

Very low concentrations of hexavalent chromium can activate the enzyme beef heart phosphodiesterase, which is regulated by calmodulin, a calcium dependent protein. MacNeil et al. suggested that the effect of chromium on this regulatory protein may be related to its toxic effects (Refs. 29 and 93).

Allergic Sensitivity. Caldas and Hicks reported that pulmonary exposure to chromate or extracts of chromium and nickel-rich fumes from SMAW of stainless steel provokes a specific tolerance to the immunologic effects of chromate in the skin (Refs. 18 and 60).

In Vitro Studies

Bacterial Assays. Biggart et al. reported that the gas phase and particulates of emissions from SMAW of mild steel are mutagenic in a modified Salmonella/Ames assay. The fumes contained both direct- and indirect-acting mutagens which induced frameshift mutations. The results suggested that fume particles become degraded at the cell surface, permitting particle components to be taken up by the bacteria (Ref. 14). Unlike the fume particles, exposure to the gas phase caused base pair substitutions but not frameshift mutations. (Ref. 13).

Ong et al. demonstrated that the SOS test, which detects gene damage in Salmonella typhimurium by a colorimetric reaction, can be adapted for use in occupational settings (Refs. 124 and 125). Stainless steel welding fumes induced a high response with the SOS test, and the effects were greater without microsomal enzymes than in their presence. The SOS test was much more sensitive to welding fumes than the Ames test. This may be related to the observations of Pedersen et al. (Ref. 137) of higher concentrations of hexavalent chromium in fumes collected by impingement than in those collected by filtration.

Mammalian Cell Studies. Two studies examined the induction of SCE by welding and cutting aerosols. Baker et al. showed that SCE activity, but not mitotic delay, was proportional to the hexavalent chromium content of the welding fume in cultured Chinese hamster lung cells (Ref. 8). The soluble fume fraction was much more active than the insoluble fraction in both the mitotic delay and SCE assays, presumably because of the greater bioavailability of the soluble components of the fume particles.

Valerio et al. showed that fumes emitted by gas-cutting during refitting operations in oil tanks cause SCE in murine bone marrow cells (Ref. 182). These fumes are known to contain high levels of benzo (a)pyrene and other polycyclic aromatic hydrocarbons which may be responsible, in part, for their mutagenic activity.

Hooftman et al. reported that the cytotoxicity of welding fume particles to bovine alveolar macrophages decreases in the following order: SMAW of stainless steel >GMAW of stainless steel >SMAW of mild steel >GMAW of mild steel (Ref. 64). Pasanen et al. (Ref. 135) found that the cytotoxicity of welding fumes to rat alveolar macrophages decreases in the following order: SMAW of stainless steel = SMAW of mild steel >GMAW of stainless steel >GMAW of mild steel. Similar to the findings of Hooftman et al., Pasanen showed that the viability of cells treated with fumes from GMAW of mild steel differed little from control cells. Removing water-soluble materials by washing with phosphate-buffered saline substantially reduced the cytotoxicity of fumes from SMAW of stainless steel but had little effect on the other fumes. The effect of washed and unwashed fumes on the release of lactic dehydrogenase followed the same pattern (Ref 135). The results of both Hooftman and Pasanen indicated that hexavalent chromium may be responsible for the cytotoxic activity of welding fumes to alveolar macrophages.

In vitro studies indicated that combined exposures to chromate and hyperbaric pressures can act synergistically to disrupt the organization of actin filaments, which are responsible for cell shape and motility and may prolong the G1 phase of the cell cycle (Refs. 68 and 173).

Conclusions

Welding emits fumes and gases which may cause adverse health effects. Emissions vary widely with the process and can be controlled to some degree by choice of method, electrode, and filler material. The use of proper ventilation, protective gear, and clothing can keep exposures to a minimum. Training programs may be needed for management and welders alike, demonstrating the economic and health advantages of reducing exposures. New training techniques, employing video tapes and computer systems, have been developed which may help the welder to a better understanding of exposures and to adjust work practices accordingly.

Unresolved questions concerning health issues persist. The most important of these concern the effects of welding on the respiratory tract and cancer rates among welders.

The Respiratory Tract. Exposure to some components of welding emissions, including ozone, nitrogen oxides, chromium, and nickel, may injure the respiratory tract. However, the relationship between welding exposures and deficits in pulmonary function, or respiratory diseases such as bronchitis and emphysema, is not well understood. The generation of injurious pollutants varies with the welding method and the metal welded. The influence of the specific chemical constituents of welding fumes on the development of pulmonary impairment is an important area that has not been adequately investigated.

Inconsistencies in reports of the chronic effects of welding fumes on the lungs may be due, in part, to studies of populations exposed to different welding materials. Therefore, studies of the incidence of respiratory diseases in welders should be carefully designed and should concern populations selected in a manner that will allow correlation of results with exposures to specific welding processes. In addition, a variety of pulmonary function tests are available. Questions have been raised as to whether some widely used tests are likely to detect the types of lesions that could potentially be produced by welding fumes. Thus, the applicability of different pulmonary function tests for research purposes or for general medical screening of welders should be further examined.

Further development of practical techniques for assessing the fibrogenicity of fumes and gases from welding operations is important for the assessment of risk associated with different electrodes and welding methods. In vivo and in vitro methods for studying the potential of occupational pollutants to cause irritation, inflammation, or fibrosis in the lung and other tissues should be explored. A comprehensive literature search to determine whether methods exist that are useful for studying welding emissions might prove fruitful.

Several research groups reported that exposure to some welding fumes can affect the function and viability of alveolar macrophages, which are important for maintaining the health of the lungs. However, since their activity may stimulate inflammatory and fibrotic processes, alveolar macrophages may also contribute to the development of chronic lung disease. Mylius and Gulvag showed that cigarette smoke and welding exposure combine to produce a marked increase in the number of alveolar macrophages in the lung (Ref. 104). Vogelmeier et al. observed a tenfold increase in the number of leukocytes obtained by bronchoalveolar lavage in a patient with recurrent metal fume fever (Ref. 185). Recognizing the potential importance of macrophage activity in the development of chronic lung disease, both of these investigators questioned whether such impressive numbers of inflammatory cells can be present in the lungs without causing damage. It may be of interest to focus more attention in the future on the effects of welding fumes on alveolar macrophages and to determine what these effects mean in terms of potential health effects.

Cancer. Fumes generated by welding stainless steel contain potentially carcinogenic chemicals, in particular hexavalent chromium and nickel compounds. The presence of these metals in fumes from some welding processes provides a major argument in support of the probability of an increased cancer risk in some welders. The mechanism of action of metal carcinogens remains unknown and is an important area for continued research. Epidemiologic studies differ widely in their conclusions concerning the cancer risk of welders. In many of the cancer epidemiology studies reviewed in this volume of Effects of Welding on Health, it is difficult to rule out nonwelding exposures among the welding populations. More carefully controlled studies of welding populations are critically needed to help understand the cancer risk of welders. Studies of large populations of welders, in which exposures to specific welding processes can be accurately determined, are necessary to provide a much needed answer to the question of whether exposures to specific fumes account for the elevated cancer risk observed among some populations of welders.

Effects of Welding on Health VII

Section One
The Exposure

Hazardous fumes, gases, and electromagnetic radiation can be generated during welding. Organic vapors may be present in welding emissions if the base metal is coated with organic products such as paints or oils. The composition of the emissions varies with the welding method. Of the most common welding processes, fume generation is lowest with gas tungsten arc welding (GTAW) and highest with shielded metal arc welding (SMAW) and flux-cored arc welding (FCAW). Ozone levels are greatest with gas metal arc welding (GMAW) of aluminum. High levels of nitrogen oxides (NOX) are produced by all processes that involve intense heat. Gas welding and plasma cutting tend to produce more NOX than other welding methods.

An industrial hygiene survey conducted in six Dutch factories showed that fume and gas exposures from plasma welding of stainless steel are comparable to the low exposures produced by GTAW, while exposures from plasma cutting of stainless steel are comparable to those produced by SMAW (Ref. 183). The Dutch occupational health standard for welding fumes (5 mg/m^3) was exceeded when plasma cutting of stainless steel was conducted without local exhaust ventilation; fume concentrations were below the standard when local exhaust was used and also during cutting under water. Hexavalent chromium was found in only two plants and its concentration never exceeded 40 ug/m^3. Total chromium concentrations did not exceed 400 ug/m^3. In no instances were the Dutch standards for nitrogen dioxide (NO_2: 5 ppm), nitric oxide (NO: 25 ppm) or nickel (1 mg/m^3) exceeded, even during the process of cutting alloys containing 30 to 35% nickel. None of the samples contained water-soluble nickel. Breathing zone levels of ozone were excessive because of the proximity of the welder's face to the torch. This problem was resolved by placing a glass shield on the work table between the workpiece and the welder.

1. Fumes

Welding fumes are composed of microscopic metal oxide particles which originate primarily from the filler metal and the electrode coat or core materials. The concentrations of fume constituents vary with the welding process, electrode, base metal, and welding parameters (voltage and current). The fume generation rate (FGR), a measure of the quantity of fume emitted per unit time, is used to compare fume emissions from different electrodes. The importance of this value for assessing relative hazards associated with different electrodes was stressed by Moreton who urged that standard methods for measurement of FGR be adopted (Ref. 103).

FGRs have been determined for a large number of electrodes used in different welding processes. Recently, Henderson et al. showed that the FGR was lower for gas shielded flux cored electrodes than for those used without gas shields (Ref. 58). Krause and Press reported that the quantity of emissions generated during resistance welding increased markedly during "splashing". Resistance welding of greased metals also increased fume formation. In the absence of splashing, a layer of grease (1.3 mg/cm^2 metal surface) increased fume formation by 30%. The emissions increased with the thickness of the grease (Ref. 85).

The effects of hyperbaric pressure on welding fume emissions is receiving attention in Northern Europe because of the importance of offshore oil operations in that area. In this regard, Bjoerseth et al. showed that for GTAW, fume production is 5 to 10 times greater at a pressure of 32 bar than at 1 bar. The ratio of manganese to iron was also greater at higher pressures, presumably due to heat distribution effects (Ref. 16).

1.1 Effects of Electrode Composition. In the interest of identifying electrodes that generate the least hazardous emissions, Olah and coworkers measured gases (ozone, hydrogen fluoride, NOX) and light and heavy elements [Fe, Cr, Cr (VI), Ni, Mn, Mo, Si, C, and others] in fumes produced by coated electrodes, flux cored hardfacing electrodes (Refs. 115, 116, 118, and 119), and by electrodes used for welding high alloy steels (Refs. 117 and 177). Analyses were conducted by neutron activation analysis, X-ray fluorescence spectroscopy, and atomic absorption spectrometry (AAS). The results showed that the exposures resulting from hardfacing are comparable to those from SMAW of high alloy steels (Ref. 116).

Tandon et al. determined 19 elements in the flux and fumes from three types of hardfacing electrodes and two types of coated, high-strength, low alloy steel electrodes. Magnetite (Fe_3O_4), potassium chromate (K_2CrO_4), calcium fluoride (CaF_2) and sodium fluoride (NaF) were the only compounds detected in the fumes. The water solubility of the fumes varied from 22.5% to 37.5% for the five electrodes. An inverse relationship between the fluoride concentration in fumes and the FGR was noted. The content of sodium and potassium in the flux was highly correlated with the ratio of water-soluble hexavalent chromium to total chromium in the fume. (Ref. 174).

In order to establish exposure guidelines, McIlwain and Neumeier of the Bureau of Mines determined FGRs and fume constituents generated by selected welding electrodes commonly used in mines. From these data, exposure indices were determined which provide a relative measure of the potential health hazards associated with emissions from the electrodes (Ref. 96).

1.2 Chromium. In 1983, Thomsen and Stern reported finding a short-lived hexavalent chromium species in freshly formed welding fumes collected by impingement in an aqueous medium (Ref. 133). GMAW fumes collected by traditional methods on dry membrane filters generally contain little hexavalent chromium, whereas Thomsen (Refs. 137 and 176) reported that GMAW fumes collected by liquid impingement contain 6 to 7% hexavalent chromium. The same investigators examined the effects of sample collection method, storage time, and analytical procedure on the concentration of hexavalent chromium in welding fumes (Ref. 137). Hexavalent chromium was reduced to the trivalent state when SMAW fume samples were stored for up to 15 days on cellulose acetate, but not on polyvinyl chloride (PVC) membrane filters, and reduction of hexavalent chromium was observed with some analytical procedures. Pedersen et al. suggested that chromium measurements made on samples collected by wet impingement may reflect actual exposures better than those made on samples collected on dry membrane filters. However, other investigators have argued that the short-lived chromium species may be an artifact resulting from chemical reactions in the wet collection medium (Ref. 133) and that much more information concerning the generation of hexavalent chromium is needed before traditional methods of fume collection can be replaced, or supplemented, by wet impingement (Ref. 46).

Hewitt and Madden (Ref. 59) used the wet impingement collection method described above (Ref. 176) during a study of the effects of various welding parameters on hexavalent chromium concentrations in fumes generated by GMAW of stainless steel. Hexavalent chromium levels were greatest with a shield gas of pure argon and decreased with increasing concentrations of carbon dioxide in the shield gas. Their finding that ozone increased the concentration of hexavalent chromium led to the conclusion that the formation of hexavalent chromium in GMAW fumes is due, in part, to reactions between trivalent chromium oxide and ozone. In his critique of this work, Gray argued that since the increase in hexavalent chromium was not proportional to the ozone concentration, ozone reactions cannot account for hexavalent chromium formation (Ref. 46).

1.3 Nickel. Nickel carbonyl is a severe pulmonary irritant and has caused cancer in animal studies. Wiseman and Chapman (Ref. 190) determined nickel carbonyl levels in emissions produced by GMAW and GTAW of Nickel 200, stainless steel, and other alloys (Inconel 600, Incoloy 800, and Monel 400). Traces (0.1 ppb) of nickel carbonyl were found in emissions from 3 of the 16 welding combinations tested, all of which involved GTAW of Nickel 200 with shield gases containing carbon monoxide (CO) or carbon dioxide (CO_2).

1.4 Analytical Methods. The utility of X-ray fluorescence analysis for determination of components in welding fumes was described by Nemcova (Ref. 106). Voitkevich and Fedorina compared different methods for determining hexavalent chromium concentrations in stainless steel welding fumes (Ref. 186). Olah and Pivoluska described a rapid method using X-ray fluorescence spectrometry to quantify chromium in fumes from welding of stainless steel. The limit of detection was 3 ug. The correlation between results obtained with X-ray fluorescence spectrometry and the diphenyl carbazide method was poor (Ref. 120).

A method employing high pressure liquid chromatography to separate beryllium, cobalt, nickel, and chromium was described by Maiti and Desai (Ref. 94). The elements were separated as beta-isopropyltropolone complexes on a reversed-phase column. The method was used to determine chromium in air samples from welding operations. The procedure did not distinguish between trivalent and hexavalent chromium.

Sawatari and Serita (Ref. 148) examined chromium species in metal fumes generated from pure chromium powders using a plasma metal sprayer. A series of leaching solutions was used to differentiate between chromium of different valences and solubilities. The procedure was rather time-consuming, but the authors indicated that it could be simplified to allow determination of the solubility and valence of chromium with only three steps (leaching with sodium carbonate, digestion of undissolved fumes with heated dilute sulfuric acid, and fusion of the undissolved portion with a boiling mixture of sulfuric acid and perchloric acid).

Chong described a method using AAS to determine concentrations of Ag, Bi, Cd, Cu, Fe, Ni, Pb, and Sn in lead- and zinc-based solders and metal alloys. The technique prevented loss of metals by co-precipitation and allowed all of these elements to be analyzed from a single sample (Ref. 21).

Ingle developed a device for automatically recording the time and duration of arcing periods during electric welding (Ref. 65). A reed relay was used as a detector of the circumferential magnetic field around the cables carrying the welding current. The relay was placed next to the cable in the direction of the field; it opened and closed as the strength of the field rose and fell. Ingle reported that the device could be used to record arcing patterns of GMAW, GTAW, and SMAW, with both direct and alternating current. The pattern obtained for the variation of ozone concentrations with arcing time during GMAW of aluminum is shown in Figure 1.

1.5 Analysis of Metals in Biological Tissue. Antilla et al. found that transmission electron microscopy and X-ray microanalysis could successfully identify some mineral and metal particles in autopsied lung tissue. However, occupational histories or additional chemical analyses were necessary for accurate identification of particles from metal grinding or welding (Ref. 6). Landsberger and Simsons showed that epithermal neutron activation analysis could be used for the nondestructive determination

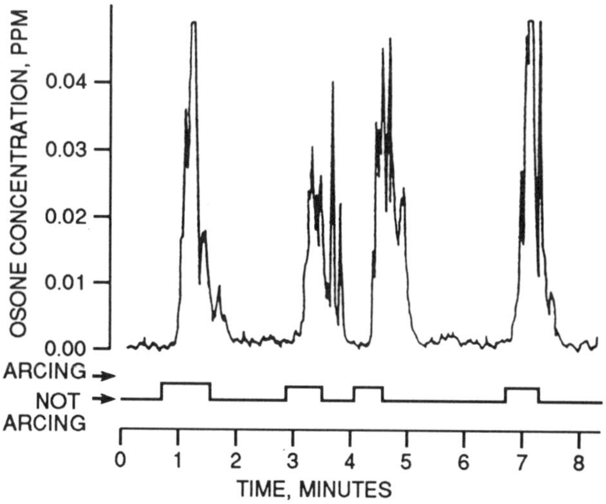

Figure 1 — Ozone Concentration and Arcing Time During GMAW of Aluminum Ingle, Ref. 65

of chromium, nickel, and zinc in biological tissues and for the determination of nickel and arsenic in urine. These techniques were used to analyze metallic deposits in lung tissue from a welder who died from cancer (Refs. 88 and 89). Baumgardt et al. described a method using AAS for measuring trace quantities of metals in biopsied lung tissue. Elevated levels of chromium, cobalt, and manganese were found in lung tissue from a welder (Ref. 10).

1.6 Particles. Microscopic particles comprise the bulk of welding fumes. The particles tend to be of respirable size (0.1 to 10 um in diameter) and are generally spherical, although the regularity of the surface may vary with the welding process and electrode. Particles may be present in fumes as single entities or as agglomerates or chains of varying lengths. The particles are chemically heterogeneous. Their composition is determined primarily by the content of the filler metal but, for SMAW, elements from the flux also contribute to the fumes.

The outer surface of particles from SMAW tends to be composed of lighter elements (e.g., Na, K, Ca, F) while heavier elements (e.g., Fe and Mn) are found in the particle core (Refs. 48 and 187). The surface and bulk elemental composition of particles from SMAW of stainless steel were determined by Grekula et al. (Ref. 48) using scanning electron microscopy (SEM) in conjunction with energy-dispersive X-ray analysis (EDXA). The surface and subsurface chemical composition (as revealed by argon-ion sputtering) differed for large and small particles. The results suggested that formation of

Table 1
Morphology and Chemistry of the Major Particle Populations in Welding Fumes

Weld Method	Particle Population	Population Diam. (um)*	Particle Size (um)	Fe	Si	Mn	Cr	Ni	Na	K	Ca	Mg	Ba	Other
GMAW/MS	Granular network	3	0.02	74	20	6								
(CO2)	Clusters of glob	1	0.1	70	13	17								
	Chains	10	0.05	82	10	8								
GMAW/SS	Granular network	up to 10	0.02	41	18	10	23	8						
(Ar/O2)	Clusters of glob	1	0.1	40	10	24	22	4						
SMAW/MS	Granular network	up to 10	0.03	86	3	1				10				
	Clusters of glob	1	0.2	22	10	7			3	27	30			1
	Blobs	-	0.5	9	2	3			7	71	7			
SMAW/SS	Granular network	Similar to		48	5	1	11	11	7	17				
	Clusters of glob	SMAW/MS		9	23	5	5	1	18	35				4
	Blobs	fume particles		1	2	1	2		29	65				
FCAW/MS	Granular network	4	0.02	75	20	5								
(Ar/O2)	Clusters of glob	1	0.1	49	11	30			6					4
	Chains	5	0.03	76	14	10								
FCAW/MS	Granular network	1	0.02	**										
(without	Clusters of glob	0.5	0.2	11	4	3				2	1	26	48	5
gas)	Chains	5	0.03	86	4	2						3	5	

SS - Stainless steel

MS - Mild steel

* - Size of particle cluster

** - "Composition varies greatly; the main component can be either Fe, Mn, Al, Mg, or Ni"

Data from Kalliomaki et al., Ref. 72

particles is partially dependent on vapor pressures. Metals with lower vapor pressures (e.g., Fe, Cr, and Ni) are enriched in the core, whereas those with higher vapor pressures (Na, K, F) condense on the particle surface.

In other studies, Grekula et al. examined the physical characteristics of particle agglomerates in fumes from SMAW, GMAW, and FCAW of mild steel (Refs. 47, 49, and 72). Particles were grouped into four classes on the basis of their morphology: clusters of globules, chains, granular networks, and large spheres (Table 1). The finest metal-rich particles formed granular networks which represented most of the particles in CO_2-shielded GMAW fumes and a minor fraction of the particle agglomerates in SMAW fumes. Particle agglomerates produced by GMAW and gas shielded FCAW of mild steel were similar in morphology. Clusters of globules predominated in fumes from SMAW and FCAW (without gas), whereas particles in GMAW and gas shielded FCAW fumes were present primarily as granular networks. Individual semitransparent particles, referred to as blobs, were unique to SMAW fumes. They were composed primarily of potassium and sodium and apparently arose from the slag. The morphology of particles from SMAW was the same for rutile and for basic electrodes. Most of the agglomerates had a characteristic chemical composition; however, the composition of individual particles within the agglomerates varied.

The size of particles in welding fumes is generally expressed as the mass median aerodynamic diameter (MMAD), which is used to estimate the likelihood of their deposition in the lungs upon inhalation. According to Ahlberg (Ref. 2), the MMAD is not completely valid for welding fume particles since they tend to be agglomerated into chains and other nonspherical masses. The equivalent mobility diameter (DME) better accounts for the sedimentation and diffusion forces to which welding agglomerates are subjected. In addition, some welding fume particles are hygroscopic and may become hydrated in the respiratory tract. This hydration could theoretically alter their deposition rates in the lungs. Ahlberg et al. determined the MMAD, the DME, and the growth in particle size as a function of relative humidity for particles in fumes from SMAW of mild steel. They concluded that hydrated particles could collapse to spheres after they enter the respiratory tract which would cause the MMAD to increase and the DME to decrease. This would result in increased sedimentation and diffusion rates causing greater particle deposition in the alveoli. The investigators concluded that the actual lung deposition is less than the deposition predicted from the MMAD for nonhygroscopic welding fumes and more than that predicted for hygroscopic welding fumes (Ref. 2).

2. Ozone

Ozone, a severe respiratory irritant, is generated by reaction of ultraviolet (UV) light with oxygen. Ozone is unstable in air and its decomposition is accelerated by metal oxide fumes. Because of this, welding processes such as SMAW and FCAW, which generate large quantities of fumes, are not usually associated with significant quantities of ozone.

Engstrom and Virtamo determined ozone levels behind the welding helmet in the breathing zones of welders from 24 welding shops (Ref. 39). A total of 118 measurements, 5 to 15 minutes in duration, were taken with several different gas-shielded welding processes. Ozone concentrations fluctuated widely during individual measurement periods. Peak levels as high as 7 ppm were recorded during GMAW of aluminum in the absence of local ventilation. Ozone concentrations were markedly reduced by the use of local ventilation and the mean concentration for all welding processes was less than the Finnish standard of 0.1 ppm. GMAW of aluminum produced the highest concentrations with a geometric mean of 0.06 ppm. The addition of nitric oxide to the shield gas was less effective in reducing ozone levels than was predicted by the laboratory studies of Sipek and Smars (Refs. 132 and 157).

3. Electromagnetic Radiation

Electromagnetic radiation in the UV, visible, and infrared (IR) regions of the spectrum is produced during welding. Overexposure to radiation in any of these categories can damage the eyes. In addition, exposure to UV radiation can injure the skin (e.g., erythema or sunburn, skin cancer). The biological effects of electromagnetic radiation vary substantially with the wavelength. Many factors affect the intensity and spectral distribution of the UV light generated by welding equipment. These include the welding process, base metal, electrode, voltage, current, and arc length.

Guidelines for determining permissible exposure times to electromagnetic radiation that relate the biological effectiveness to the wavelength (effective irradiance) have been established by the American Conference of Governmental Industrial Hygienists (ACGIH) (Ref. 1). Using these guidelines, Okuno estimated permissible exposure times at 1 meter from the source for the UV light generated by various welding processes (Ref. 114). The results are shown in Table 2.

The permissible exposure times varied from several seconds to one minute for most of the welding processes studied. For each welding process, the UV radiation increased with the diameter of the wire, arc current, and voltage. As expected, the UV hazard was greater with GMAW than with SMAW. Okuno concluded that permissible exposure times could readily be exceeded by welders, as well as by other persons in the vicinity.

Using the same welding conditions, Okuno measured the blue light radiation at a distance of 1 meter from the arc (Ref. 113). No correlation was found between the blue light irradiance and the diameter of the wire, arc current, or voltage. The greatest blue light effective radiance (2.4 W/cm^2) was found for CO_2-shielded GMAW of mild steel, for which a permissible daily exposure time of 42 seconds was estimated. In similar work, Salsi and Barlier measured the spectral emissions from UV to far IR (200 to 3000 nm) generated by 47 diverse coated electrodes. The ACGIH guidelines for UV, visible light, and IR radiation were used to calculate daily permissible exposure levels and to evaluate the shade numbers of filters to be used with the electrodes studied (Ref. 9).

Proper precautions to protect the eye must be exercised during laser welding, as the cornea and retina can be permanently damaged by exposure to laser beams. The extent of the hazard varies with the characteristics of the laser beam and, thus, selection of protective eyewear must address the specific laser process used. According to Winburn (Ref. 189), goggles supplied or recommended by laser manufacturers often provide more protection than needed and, in so doing, do not promote optimal visibility. For example, colored plastic goggles are

Table 2
Permissible Exposure for UV Light From Diverse Welding Processes

Process	Metal	Diam. (mm)	Current/ Voltage	Shield Gas	Permissible Exposure Time Per Day (sec)
SMAW ilmenite	MS	2.6	70A 20V AC	N/A	300
SMAW ilmenite	MS	4.0	170A 30V AC	N/A	51
SMAW ilmenite	MS	8.0	400A 36V AC	N/A	33
SMAW low hydrogen	MS	4.0	170A 23V AC	N/A	36
SMAW low hydrogen	MS	8.0	400A 32V AC	N/A	17
GMAW solid wire	MS	1.2	150A 23V DC(+)	CO2	23
GMAW	MS	1.2	150A 215V DC(+)	5Ar-CO2	14
GMAW solid wire	MS	1.6	300A 35V DC(+)	CO2	16
GMAW	MS	1.6	300A 33V DC(+)	5Ar-CO2	4.9
GMAW	SS	1.2	270A 35V DC(+)	CO2	21
GMAW	Aluminum	1.6	250A 27V DC(+)	Argon	31
Self-shielded	MS	3.2	400A 25V AC	N/A	250

N/A - Not applicable
SS - Stainless steel
MS - Mild steel
Self-shielded - 'Self-shielded arc welding"
Data from Okuno, Refs. 113 and 114

sometimes provided by manufacturers for use with CO_2 lasers. Wavelengths produced by these lasers are readily absorbed by almost all transparent materials and any clear goggles would serve as efficient protective devices without reducing visual acuity.

4. Production Coatings

Toxic substances may be released when welding metals coated with paints, primers, greases, and other materials. Heikila et al. (Ref. 53) showed that phenanthrene, anthracene, pyrene, benzo(a)pyrene and other polycyclic aromatic hydrocarbons are released during SMAW of railroad ties. Benzo (a)pyrene, a potential human carcinogen, was measured in concentrations as high as 0.64 ug/m^3 in the vicinity of the welders.

5. Noise

Impulse noise may be more harmful to hearing than continuous steady state noise of the same energy level (Ref. 133). The impulse noise level may be defined as the difference between the peak value and the root mean square (RMS) of the slow time-weighted value; this difference is also referred to as the crest factor. Noise is considered to be impulsive if the difference between peak levels and the slow time-weighted value is greater than 15 dB.

Starck and Pekkarinen described a telemetric procedure for measurement of industrial impulse noise (Ref. 168). The method provides data for both equivalent noise level and impulsiveness. Measurements were made with both a stationary sampling device, which provides only a limited estimate of a worker's noise exposure, and with a miniature microphone placed in the worker's ear. Earlier tests, performed at a shipyard plate welding workshop, indicated that grinding and carbon arc work produce minimally impulsive noise, whereas welding processes, especially GMAW, generated high impulse noise levels (Ref. 133). A more recent study (Ref. 168) was conducted in a shipyard assembly hall. The impulsiveness for different processes increased in the following order: grinding < carbon arc gouging < chiseling < welding (process not specified) < GMAW. Noise generated by GMAW had the highest impulsiveness and relatively low energy equivalent levels, while that generated by gouging and grinding had the lowest impulsiveness and the highest equivalent levels.

Impulse noise generated by different processes was analyzed with a transient recorder; the results for grinding, chiseling, and GMAW are shown in Figure 2. Chiseling generated recurrent impulse noise, with peaks at intervals of about 20 milliseconds, whereas grinding produced random, nonimpulsive noise. GMAW produced short impulses that occurred randomly at intervals of tens of milliseconds. The impulsive noise peaks from GMAW had average equivalent levels of 90 dB and maximum levels of 120 dB. While technical measurements proved noise from GMAW to be highly impulsive, noise generated by GMAW does not appear to be impulsive to the human ear because the peaks are too short to be transmitted through the auditory nerve to the brain.

Tests designed to determine whether hearing protectors effectively attenuate impulsiveness (Ref. 138), were conducted at 21 different workplaces. The average equivalent noise varied between 85 and 98 dB outside the earmuffs and between 69 and 83 dB inside the earmuffs. The average attenuation of equivalent noise levels was 17 dB. Earmuffs reduced the equivalent noise levels below 85 dB(A) in 90% of the measurements. The percent of the total measurement time during which the peak sound level exceeded the RMS level by at least 15 dB, was 1.6% outside the earmuffs and 0.8% inside the earmuffs. Thus, earmuffs were effective in reducing impulse noise. It was noted that the age of the earmuff and problems with individual fit strongly affect the efficiency of these hearing protectors.

6. Hygiene and Work Practices

6.1 Accidents. Accidents are among the most important health risks for welders. A survey in Washington State of accidental fatalities that occurred in the construction industry between 1973 and 1983 indicated that welders ranked third among occupational groups. (Ref. 17). Accidents are generally considered to be the primary health risk associated with the use of robots. Currently, welding operations represent the most common use of robots in industry. Although published statistics on robot-related accident rates are limited, a study performed in Japan indicated that the majority of accidents occur during teaching, testing, and maintenance. Accidental injuries and deaths mainly involve impact, pinch, or release of an object from the robot's arm.

A common cause of accidents is pushing the wrong button in the robot controls. Other accidents are caused by improper programming of robot movement. Some accidents occur because of failure or difficulties in perceiving or anticipating the movement of the robot arm. Few standards exist for controlling the maximum speed of arm movements. The robot arm should move slowly enough to give sufficient time for operator reaction in case of emergency. However, arm movement should be fast enough to be readily perceptible by the operator or other personnel in the area. Pauses in work operations should be short enough to prevent the mistaken perception that robot operation has stopped (Refs. 56, 57, and 134).

Some robots are being equipped with auxiliary safety devices which monitor robot operations. If persons or extraneous objects intruding in the robot work environment are detected, the operation of the robot is automatically terminated. It has been suggested that, in addition to this "watchdog" device, personnel who must function within the robot enclosure should wear a device that transmits a signal to the robot's safety sensor, indicating human presence (Ref. 57).

De Nevers described an accident that occurred in a facility where solid rocket boosters from the space shuttle program are cleaned and inspected (Ref. 30). The welder was in a basket suspended from an overhead crane inside a vapor degreaser pit filled with 1,1,1-trichloroethane vapors. When he started welding, he noticed "long hydrocarbon-type flames" around the arc. He resumed welding after allegedly being informed by the facility engineer that it was safe to proceed because 1,1,1-trichloroethane would not burn. During the welding operation, a ball of flame developed near his chest and rapidly expanded; the welder fell into the pit as a huge fireball erupted. His body was severely burned, but it was not clear whether his death was due to burns or to asphyxiation by 1,1,1-trichloroethane fumes. Although 1,1,1-trichloroethane has no flash point, and the liquid solvent will not burn, according to De Nevers, the vapors can burn in the

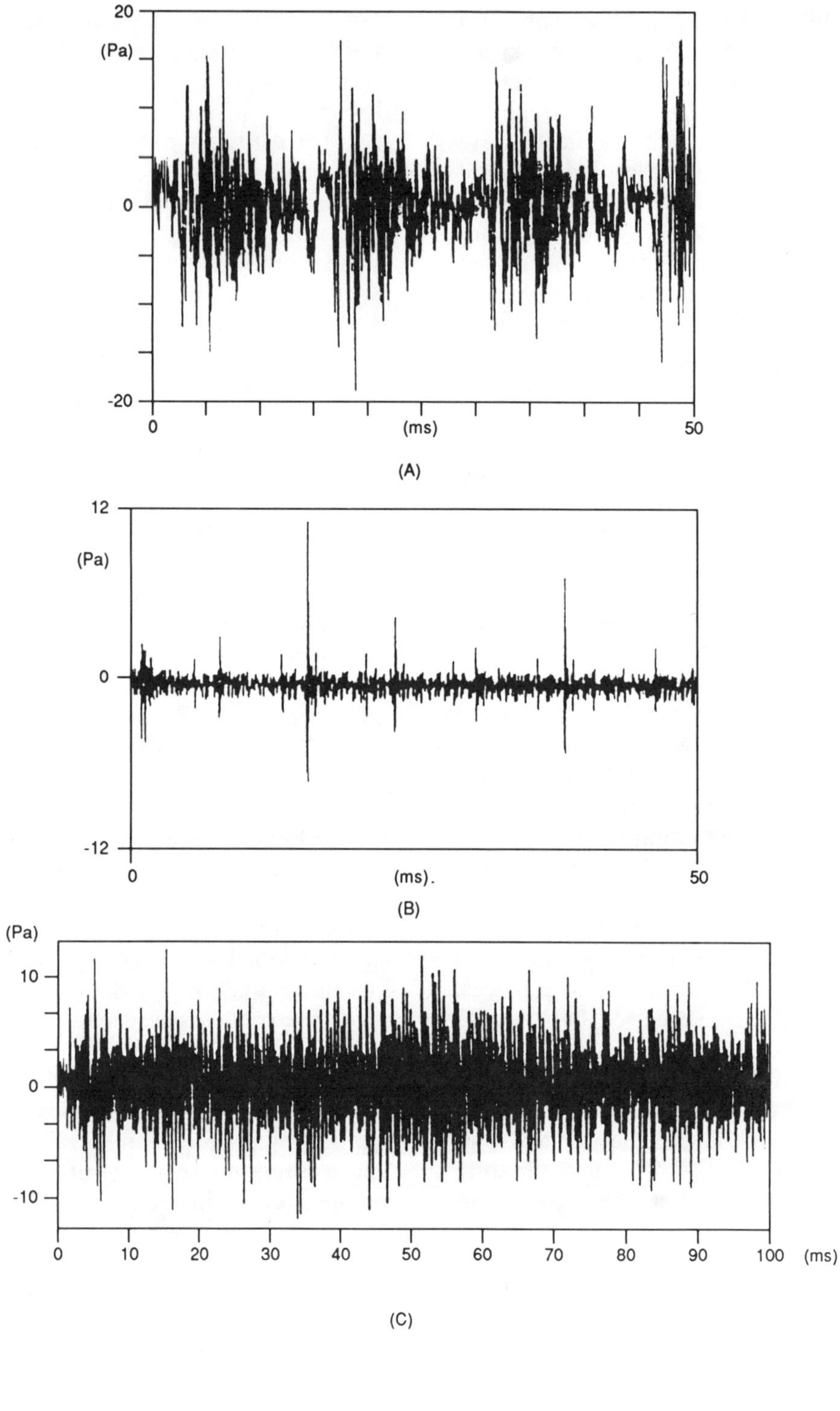

Figure 2 — The A-Weighted Sound Pressure Measured With a Stationary Microphone and Transient Recorder for (A) Chiseling (B) GMAW, and (C) Grinding. Starck and Pekkarinen, Ref. 168

absence of the liquid phase. Commercial formulations frequently contain corrosion inhibitors which increase the flammability of the solvent. Tests of the 1,1,1-trichloroethane remaining after the accident indicated that it contained twice the quantity of corrosion inhibitors normally used, which may have contributed to the flammability of the vapors.

6.2 Clothing. To determine their effectiveness in protecting welders from actinic UV radiation, Sliney et al. measured the attenuation of UV by fabrics commonly used in industrial work clothing (Refs. 12 and 163). The direct and diffuse transmission of UV light were measured separately for 45 different fabrics at five wavelengths between 220 and 340 nm. Most fabrics transmitted less than 0.01% of the incident light, which was considered to be adequate for protection against UV radiation from the welding arc. However, two lightweight fabrics had transmission values close to 1%, which permitted transmission of sufficient UV light to exceed current occupational exposure limits.

The attenuation of transmitted light varied little with the wavelength, which indicated that light passed through openings between the fibers. Laundering decreased the UV transmission through all but one of the fabrics. This was thought to be due to shrinkage of the fabrics with a consequent decrease in the size of the openings between fibers. (The effect of laundering on UV attenuation for several fabrics is shown in Figure 3.) This study indicated that most fabrics used in standard work clothing provide adequate attenuation of the UV radiation generated by welding arcs; thick, dense fabrics are better than lightweight materials. The investigators recommended the use of long-sleeved clothes and gloves made of thick, dense material with an optical density of 4 during welding operations with currents up to 400 A.

Although work clothes should be impermeable to UV and should provide some protection against burns, they should also allow dissipation of sweat and body heat. A study by Beckett et al. demonstrated that vapor-barrier garments in the workplace can cause heat stress (Ref. 11). Plumbers, insulators, and welders, wearing vapor-barrier suits for protection against asbestos, were examined while working. One subject had a heart rate of 164 beats per minute and an oral temperature of 100.6°F while wearing a vapor-barrier suit over conventional clothing. When wearing conventional clothing only, his heart rate was 112 beats per minute, and his oral temperature was 98.3°F. The authors noted that when heavy muscular work is performed in a warm environment, impairment of the dissipation of sweat by vapor-barrier garments could result in significantly higher body temperatures with a high risk for heat-related disorders, including heat stroke.

6.3 Protective Gear. Jakubcik (Ref. 66) described a lightweight (16 grams), maintenance-free, disposable respirator used in a steel works in Czechoslovakia to provide protection when welding high alloy steels with electrodes containing chromium, nickel, and manganese. The RVD-F respirator, developed in Carlsbad, covers the mouth and nose, fits readily under the welder's mask, and has an expected lifetime of about one week. Particles retained on the dust mask were analyzed by SEM and EDXA. Particles as small as 0.2 um were retained by the mask and the elemental composition varied with the size and shape of the particle.

Several different methods for extracting welding fumes from the work area (Refs. 67 and 145) and from hyperbaric chambers (Ref. 35) were published. Olander developed equations that related plume height and shape to temperature gradients. With this work he demonstrated that local exhaust ventilation provides the most efficient removal of fumes from the welder's work area (Ref. 121).

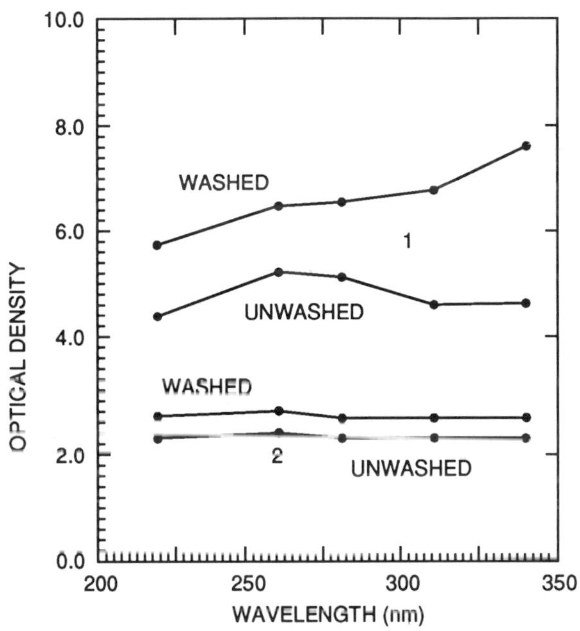

Note: Fabrics were washed ten times under standard conditions at 140°C. Lines 1 and 2 represent results obtained with different fabrics.

Figure 3 — Effect of Laundering on the UV Transmission Through Fabrics Used in Work Clothing Sliney et al., Ref. 163

6.4 Training. To facilitate the identification of work practices that can lead to excessive exposures to pollutants, Rosen and Lundstrom coupled video taping of workers and direct-reading instrumentation of breathing zone exposures (Ref. 146). The resultant video tape of work practices, with a superimposed presentation of real-time exposure data, was used to evaluate the relationship between exposures to fumes, dusts, or solvents with individual work patterns and to train persons who work with hazardous substances. The authors suggested that this method could be augmented to include IR radiation and light scattering techniques to enable visualization of gaseous emissions and particulates, respectively.

Section Two Effects of Welding on Human Health

Welding generates physical and chemical agents, including gases, fumes, radiation, and noise which can produce adverse health effects. If welders are not properly protected, exposure to welding emissions may harm the lungs, eyes, ears, and skin. Overexposure to fumes containing certain nonferrous metals or metal alloys may also injure the nervous system, kidney, and possibly other internal organs. While a great deal of research has been performed on the health effects of welding, much is still unknown about how the complex emissions generated by welding affect the body. A problem inherent in research on the health effects of welding is the great variability in welding processes and in working conditions. This makes it difficult to conduct epidemiologic studies on homogeneous populations of sufficient size to permit statistical analysis. Understanding health effects is essential to the design of protective clothing and equipment and to minimizing hazardous emissions from welding processes. Described in this chapter are health reports which appeared in the published literature from January 1986 through December 1987.

7. Respiratory Tract

Welding emissions may contain irritant particles and gases such as nitric oxide (NO), nitrogen dioxide (NO_2), hydrogen chloride (HCl), ozone, and phosgene which can elicit an inflammatory response in the lung. The intensity of the response is dependent on the dose, the physical/chemical properties of the irritant, and the overall health status of the exposed person. Macrophages migrate into the lung in response to the presence of irritants. These cells proliferate mediators, causing inflammation of respiratory tissues which may result in cough, chest tightness and pain. Severe respiratory irritants, such as ozone, phosgene, and cadmium, can cause accumulation of fluids in the lung, followed by chemical pneumonitis or pulmonary edema within hours after exposure. Chronic exposure to low levels of some irritants may result in loss of pulmonary function, emphysema, and bronchitis. The association between exposure to welding fumes and the development of chronic lung diseases, such as bronchitis, emphysema or lung cancer, has not been clearly established.

Some of the particulate matter that enters the lungs is engulfed by macrophages (phagocytosis). Many of the free particles and particle-laden macrophages are removed from the lungs by coughing and the natural movement of fluids lining the air passages. The remainder of the particles may be deposited in the lining of the lungs where they can become surrounded by fibrous material. These pockets of particles appear as opacities on chest X-rays, a condition known as pneumoconiosis. The term "siderosis" refers specifically to deposits of iron in the lungs. Although some particles (e.g., iron and aluminum oxide) are relatively inert and can remain in the lungs for years without adverse effects, combined exposure to these metals with other agents may lead to fibrotic processes and ventilatory impairment (Ref. 31). Other particles such as asbestos and quartz are strong irritants and can induce inflammatory processes and, eventually, scarring of the lungs. Inhalation of sufficient quantities of such materials can lead to extensive deposition of fibrous or scar tissue, appearing in X-rays as massive or coalescent opacities instead of isolated spots. Such scarring can cause loss of elasticity in the lung and reduced pulmonary function.

7.1 Alveolar Macrophages. Some of the lung damage that results from chronic exposure to low levels of irritants is thought to be caused by proteolytic enzymes released from macrophages at the site of inflammation. These enzymes hydrolyze proteins and hence may damage cells and tissues in their vicinity. Antiproteinases, molecules that inhibit the action of proteolytic enzymes, are normally present in the lungs and are thought to provide some protection against the destructive activity of proteolytic enzymes. Based on the assumption that antiproteinases are necessary for protection against the deleterious effects of inflammatory processes in the lung, Pierre et al. (Ref. 140) examined the interactions of welding fumes with two lung antiproteinases alpha-1-proteinase inhibitor and bronchial inhibitor. They found that welding fume particles bound reversibly to these antiproteinases and prevented them from inhibiting the activity of proteolytic enzymes.

The number of macrophages in the lungs may increase substantially in response to inhalation of irritants. Mylius and Gullvag used the number of alveolar macrophages in sputum as a measure of the inflammatory response elicited by foreign materials in the lung (Ref. 104). They showed that sputum from coke plant workers, aluminum plant potroom workers, asphalt workers, and shipyard welders had more alveolar macrophages than controls. Smoking and occupational exposures combined to give a marked synergistic effect (Figure 4). The investigators reasoned that the continuous release of proteolytic enzymes from alveolar macrophages in chronically exposed persons may represent a hazard to lung tissue.

7.2 Estimation of Retained Particles in the Lungs. Some of the particles in welding fumes contain Fe_3O_4 or other minerals (e.g., maghemite or gamma Fe_2O_3, and some sulfides of iron, nickel and cobalt) that can be magnetized. Magnetopneumography, developed in 1973 by Cohen (Ref. 24), uses this property to estimate the burden of occupational dusts in the respiratory tract. A magnetic field is applied to the chest which magnetizes and rotates the particles into a common alignment. When the external magnetic field is removed, the remanent field of the magnetized particles is determined and used as a measure of the amount of dust in the lungs. The principles involved in this technique, its applicability, advantages, and limitations were reviewed by Lippmann (Ref. 91).

A difficulty associated with the application of magnetopneumography to welding is that the magnetic properties of fumes produced by different processes vary. For example, fumes from GMAW of stainless steel have a high specific remanence while that of fumes from SMAW of stainless steel is low. If a welder is exposed to both types of fumes, the remanent field generated by GMAW will dominate, and the dust burden from SMAW fumes will be underestimated. To overcome this problem, Kalliomaki et al. (Ref. 75) developed a system which measures the pulsed coercive force. The coercive force is the reverse magnetic field which must be applied to just cancel the remanent field. It has a characteristic value for individual industrial aerosols and can be used to discriminate between remanent fields produced by different fumes.

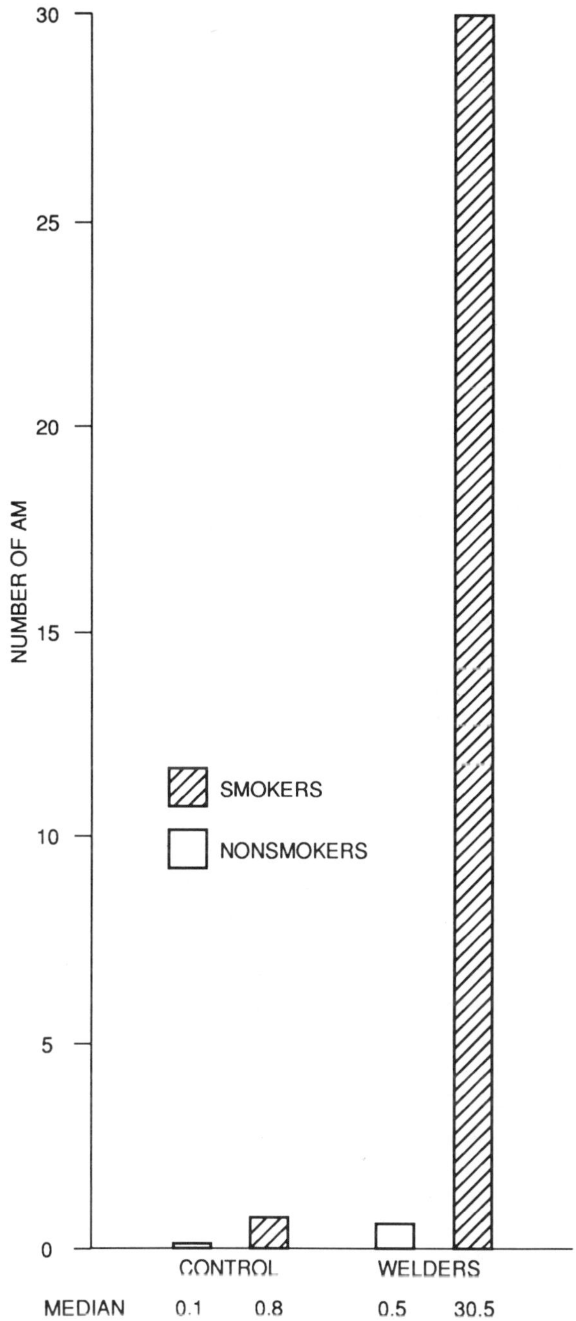

Figure 4 — Alveolar Macrophage (AM) Counts in 9 Nonsmoking and 11 Smoking Welders at a Ship-Building Yard. Mylius and Gullvag, Ref. 104

An innovation introduced by Drenck and Stern (Ref. 36) combined AC susceptibility bridge magnetopneumography with a DC method similar to that originally designed by Kalliomaki et al. The system provides information about the total content of ferromagnetic, paramagnetic, and diamagnetic material. Using this system, Stern et al. (Ref.

172) characterized the magnetic properties of fumes from a variety of welding processes and other industrial aerosols. X-ray diffraction and Mossbauer spectroscopy indicated that substituted magnetite is the primary magnetic component in freshly formed welding fumes, whereas a less substituted form of magnetite appeared to be the predominant magnetic component in inorganic dust obtained from ashed lungs of deceased welders. Magnetopneumographic studies on live welders also provided limited evidence for in vivo changes of welding fume particles to the magnetic constituent magnetite. The investigators concluded that these changes may result from loss, with time, of the more soluble components of welding fume particles.

A preliminary report by Kalliomaki et al. (Ref. 76) also suggested that the remanent magnetic field of welding particles in the lungs may not be stable. Four different types of welding fumes were introduced into the respiratory tract of rats by intratracheal instillation and the specific magnetic moment (magnetic moment divided by iron concentration) was followed for the succeeding 106 days. The specific magnetic moment of welding fumes generated by SMAW of mild steel and GMAW of stainless steel were stable, while the magnetic moment of welding fumes generated by SMAW of stainless steel and GMAW of mild steel decreased by about 50% during the 106-day follow-up period. Little change was seen in the remanent magnetism of welding fumes during a 55-day incubation in hot water (70°C) or physiological saline (70°C). On the basis of these data, it was postulated that the instability of the remanent moment of some welding fumes could be due to the slow transformation of gamma-Fe_2O_3 to alpha-Fe_2O_3 or to differences in solubility of particulate material. In addition, differences in clearance rate among different sized particles could contribute to changes in the specific magnetic moment.

7.3 Pulmonary Function and Bronchitis. Pulmonary function tests are used to detect disease processes, such as fibrosis and emphysema, which restrict lung expansion or reduce pulmonary elasticity, or both. These tests measure the air volume which can be inhaled or expelled either forcefully or under normal breathing conditions. Pulmonary function tests are often used to monitor industrial workers for damage to the respiratory tract. However, these measurements are not always sensitive to early changes in the lungs, and irreversible damage may occur before noticeable reduction in pulmonary function is detected.

Disparities have appeared in the literature concerning the effects of welding on pulmonary function. It has been suggested that differences in results of pulmonary function studies in welders may stem from the use of tests that measure those parameters that are least likely to be affected by welding (Refs. 74 and 81). Upon inhalation, the minute particles in welding fumes would tend to be deposited in the distal airways and alveolar spaces of the lung, where most pathological changes would be expected. In this regard, it has been argued (Refs. 74 and 81) that pulmonary function tests which measure impairment in the small airways e.g., forced expiratory flow from 25 to 75% (FEF25-75) and FEF75-85 are more appropriate for measuring pulmonary function in welders than are the more conventional spirometric indices [e.g., vital capacity (VC), forced expiratory value in the first second (FEV1) and FEV%]. Additional inaccuracies in the results of pulmonary function studies in welders may result from reports of respiratory symptoms among welders with confounding exposures such as asbestos or cigarette smoke. Finally, it is not known how fumes from different welding processes vary with respect to chronic effects on the lungs and, thus, comparisons of welders working with different processes may be misleading.

In their survey of welders from nine Danish shipyards, Lyngenbo et al. (Ref. 92) examined the incidence of bronchitis and upper respiratory tract symptoms (colds, sore throats, hoarseness) in 2660 welders and 881 electricians. The welders were divided into two groups: 1516 had high exposures (greater than 10 hours per week) and the rest had lower exposures (10 hours per week or less). Most of the welders had performed SMAW of mild steel for at least 20 years, and both groups were comparable in respect to asbestos exposure and tobacco use. Chronic bronchitis and upper respiratory tract symptoms occurred significantly more often in welders than in engineers, and the frequency of respiratory symptoms was related to the extent of welding exposure. There was an additive relationship between smoking and welding in regard to development of respiratory symptoms. The investigators estimated that 8 hours of welding is equivalent to smoking 25 cigarettes.

Pulmonary function tests (Ref. 92) and magnetopneumographic measurements (Ref. 171) were performed on 74 welders from the high-exposure group who had never smoked and had never worked with asbestos or other materials known to produce lung damage. The welders had significantly impaired values for VC, forced vital capacity (FVC), total lung capacity (TLC), diffusion capacity (TCO), peak expiratory flow (PEF), FEV1, and the maximum expiratory flow/75% (MEF75) relative to controls. There was no association between the quantity of metal contaminants in the lung and the incidence of chronic bronchitis. The major conclusions drawn from this study were that (1) SMA welders have a significantly elevated incidence of upper and lower respiratory tract symptoms; (2) the incidence of respiratory tract symptoms correlates positively with the extent of exposure; and (3) welders with high levels of exposure and no history of tobacco use have an elevated incidence of deficits in lung function (Ref. 92).

Kilburn et al. (Refs. 81 and 82) examined lung function and respiratory symptoms in 148 welders from a Los Angeles shipyard. The study group consisted of 73 nonsmokers with 11.3 years' welding experience, 18 ex-smokers who had welded for 16.5 years, and 57 current smokers who had welded for 10.5 years. The frequency of respiratory illness, shortness of breath, and chest pain or heaviness during exercise was more prevalent in all three welding groups than in controls. The prevalence of bronchitis was lowest in nonsmoking welders, but was still substantially higher in this group than in a control population stratified according to tobacco use. Of the spirometric measurements, the major finding was a 20% reduction in FVC and FEV1 in welders who smoked. Nonsmokers had significant reductions in FVC (4.3%), FEV1 (9%), FEF25-75 (4.4%) and FEF75-85 (14.4%). The reduction of these parameters in nonsmoking welders was less than half that in cigarette smokers, suggesting a synergism between welding fumes and cigarette smoke. The investigators recognized that prior exposure to asbestos may have confounded pulmonary function data obtained in this study.

As a continuation of their studies of the long-term effects of welding exposure (Ref. 133), Spacilova and Hykes followed lung function and chest X-ray patterns in 74 welders for 8 years (Ref. 166). In nonsmoking welders, lung function and chest X-rays were normal at the start of the study period and showed no statistically significant changes after 8 years. Welders who smoked showed initial stages of respiratory obstruction and even more frequently suffered from chronic bronchitis. The X-ray patterns progressed in 24% of the smokers versus 4% of nonsmokers and regressed in 6% of the smokers and 14% of nonsmokers.

Similar results were obtained when the welders who ceased welding at the start of the 8-year follow-up period were studied independently from those who continued welding. As for the entire cohort, there was no statistical difference in pulmonary function or in the degree of nodulation evident in chest X-rays at the beginning and end of the follow-up period. Spacilova and Hykes concluded that X-ray findings of nodular opacities in arc welders are not necessarily indicative of a pathological condition and that welding exposures alone do not lead to serious pulmonary impairment. However, smoking in combination with welding presents a much more severe hazard than welding alone (Ref. 166).

In another study by these investigators, the general health status of 73 of the welders was compared with that of 52 persons with a history of occupational exposure to silica. The incidence of chronic bronchitis and ischemic heart disease did not differ significantly between the two groups. In addition, there were no significant differences in the incidence of hypertension, obesity, or diabetes (Ref. 165).

Kalliomaki et al. (Ref. 74) followed changes over a 6-year period in pulmonary function and the pulmonary dust load of shipyard welders. The 47 SMA welders participating in the study worked mostly with mild steel. They were divided into four groups according to their total welding experience. (Active welders had approximately 4, 18, or 26 years welding experience, while retired welders had approximately 25 years welding experience.) In general, lung function values of retired welders were significantly lower than normal while those of younger welders were within normal limits. The mean VC of retired welders was significantly lower than that of men who had welded for about 4 years while the mean carbon monoxide diffusion capacity of retired welders was significantly lower than that of all active welders. Values for CV did not differ between any of the groups.

Among welders with 4 years experience at the start of the study, the quantity of metals retained in the respiratory tract increased linearly with exposure time during the first 5 years, with an average of 70 mg dust being retained per year. The quantity of lung contaminants was greatest in persons who had welded for 18 to 26 years. The lung contamination among these welders decreased slowly during the 6-year follow-up period, reflecting engineering improvements in the workplace. In retired welders, metals were cleared at a mean rate of 18% per year.

Schneider et al. previously reported that the incidence of bronchitis and nodular lung opacities was significantly greater in 433 SMA welders of mild steel than in 421 nonwelding controls from the same factory (Ref. 133). In that report, deficits in FEV were related to the duration of welding experience. More recently, these values were related to welding fume exposure as measured by personal air monitoring. The prevalence of impaired FEV1/VC, decrease in FEV1 following acetylcholine challenge, chronic cough, X-ray opacities, and pain during exercise increased with exposure to welding fumes. Manganese levels in hair were positively associated with spirometric test results, nasal sinus opacities, and pain during exercise (Ref. 149).

The prevalence rates of large airway obstructions (FEV1) and pneumoconiosis were the same among 158 GTA and GMA welders of stainless steel and aluminum and 159 controls. The welders had a significantly higher rate of disturbed ventilatory distribution (V2/V1) and chronic phlegm production. The frequency of impaired FEV1(VC,V2)V1, MEF50, and chronic bronchitis varied positively with chromium levels in hair (Ref. 149).

Schneider et al. also performed a cross sectional study of shipyard welders who had worked with mild steel for about 18 years. Siderosis was evident in chest X-rays from 15% of the welders and 8% of nonwelding controls who worked at the same shipyards (Ref. 150). More extensive clinical tests, including spirometry, flow-volume curves, body plethysmography, single-breath diffusion capacity, pulmonary compliance and ergooxytensiometry, were performed with 56 of the 64 welders with siderosis. The results indicated that siderosis is associated with a slight deterioration of pulmonary function (Refs. 150 and 151).

Krasnyuk et al. (Ref. 84) examined the effects of exposure to high and low alloy welding fumes on the respiratory tract and cardiovascular system. The study group consisted of 500 welders, about half of whom were exposed to fumes containing manganese, chromium, and nickel; 200 engineers from the same company served as controls. Chronic bronchitis and pneumosclerosis, the major pathologic respiratory conditions noted, were more frequent in welders than in controls. Chronic bronchitis was twice as prevalent in low alloy steel welders while pneumosclerosis and pneumoconiosis occurred more frequently in welders of high alloy steel. The VC was decreased in welders with overt pathologic pulmonary processes. The degree of this impairment increased with the length of employment in welding. Restrictive disturbances were more severe in welders of high alloy steel and smoking exacerbated impairment in pulmonary function. Hypertension and electrocardiographic changes were more common in high alloy than in low alloy steel welders.

Cotes and El-Gamal (Ref. 26) conducted clinical and spirometric tests on 596 welders from a shipyard in northeast England. Welding exposures were associated with siderosis, but not respiratory impairment. No ill effects were associated with burning. Welding aggravated chronic bronchitis in smokers, while both welding and burning reduced the FEV1 and increased the index of airflow obstruction in smokers. The investigators concluded that fumes from welding and burning interact with tobacco smoke to cause respiratory symptoms and some impairment of lung function. Tests conducted with 345 of the younger shipyard workers (ages 23 to 47 years) showed that exercise ventilation increased in association with welding or caulker/burning in confined or semiconfined spaces; age and smoking enhanced this effect. The investigators concluded that welding in confined or semiconfined spaces causes respiratory symptoms, some impairment of lung function, and an increased ventilatory cost of exercise in smokers (Ref. 26).

Ulrich et al. (Ref. 179) compared chest X-rays and pulmonary function in 79 persons exposed to chromium during welding of austenitic steel, 109 current welders of mild steel, and 93 former welders of mild steel. Austenitic steel welders were exposed to an average total chromium concentration of 0.152 mg/m^3. The incidence of chronic bronchitis was 53% for welders of austenitic steel, 62% for welders of mild steel, and 36% for retired mild steel welders. The prevalence of positive X-ray findings and results of pulmonary function tests (FVC and FEV1 were normal; MEF50 was reduced by 13 to 17%) were similar in all welders.

Moravik (Ref. 102) examined lung function and the incidence of respiratory symptoms in a cohort of 121 welders in Czechoslovakia. About 70% of the welders worked with CO_2-shielded GMAW while the remainder worked with SMAW. Welders took an average of 15 days of sick leave per year as compared with an average of 4 days per year taken by the 42 office worker controls. The incidence of bronchitis was significantly greater in welders who smoked than in those who did not. In addition, nine of the nonsmoking welders had bronchitis while none of the nonsmoking controls had this disorder. Deficits in pulmonary function increased with the duration of welding experience; the relationship between pulmonary function and smoking was not discussed. The incidence of pneumoconiosis was directly related to the duration of welding experience.

Oleru and Ademiluyi (Ref. 122) noted changes in pulmonary function in 16 welders in a metal cutting and welding factory in Nigeria. Dust levels were estimated to be above 5 mg/m^3 in welding areas. The welders, as well as workers in the same factory with other exposures (13 paint dippers, 7 metal cutters, 18 aluminum workers), had reductions in FEV1 and FVC relative to controls. Acute 40-hour changes in FEV1 and FVC were observed in eight welders when measurements taken at the beginning and the end of the work week were compared. Acute 8 hour, but not 24 hour, changes in peak expiratory flow rate were also observed in the welders.

Dolan and Dolan (Ref. 34) studied the prevalence of pulmonary disease in 104 welders and pipefitters with 20 or more years experience in diverse industries from three population centers in California. Pulmonary function and chest X-rays were normal in only 12% and 16%, respectively, of the welders. About 95% of the welders had worked with or around asbestos. The prevalence of asbestosis precluded the evaluation of the effects of welding fumes on pulmonary function.

Desmeules et al. reviewed medical histories of 68 welders with siderosis (Ref. 32). Mean pulmonary function values were normal but 31 had chronic bronchitis. Ten of 17 persons with inspiratory rales reported exposures to dusts other than welding aerosols. The progression of lesions was followed in 27 welders with siderosis. Follow-up X-ray examinations were performed a mean of 8.8 years after initial diagnosis. Lesions regressed in 14 cases, 5 cases remained stable, and opacities increased in 8 cases. Exposure to nonwelding dusts was reported for 6 of 13 cases with stable or progressing opacities, but in none with regressing lesions (Ref. 33). It was concluded that dust exposures other than welding may be responsible for progression of pulmonary abnormalities in welders with siderosis.

Based on an extensive review of the literature, Zober concluded that siderosis is a benign condition and pulmonary fibrosis is a rare disease in welders (Ref. 192). His analysis indicated that there is no statistical evidence for an association between obstructive lung disease and welding. Accordingly, for purposes of worker compensation in West Germany, fibrotic lung conditions can be considered as occupational diseases only if there are proven exposures to known fibrogenic substances such as asbestos, silica, or certain metals (chromium, beryllium, cadmium).

In summary, the majority of the 13 studies described above found deficits in lung function or excess chronic bronchitis among welders, or both. The reports differed on the contribution that tobacco smoke made to the development of these impairments. In one study (Ref. 34), there was a high prevalence of asbestosis among welders, while in another study the number of welders was small and the potential for nonwelding exposures appeared to be high (Ref. 122).

Of the remaining eleven studies, five found that nonsmoking welders have an excess risk for bronchitis (Refs. 80, 81, 84, 92, 102, and 149) while two (Refs. 26 and 166) found a high incidence of bronchitis only among welders who smoked. Two studies found deficits in pulmonary function in smoking welders, but none in welders who did not use tobacco products (Refs. 26 and 166). Three other studies found an elevated incidence of pulmonary deficits in both smoking and nonsmoking welders. In a small cohort of shipyard welders, Kalliomaki (Ref. 74) found deficits in pulmonary function only among retired shipyard welders and not among the younger, active welders. Finally, the investigators from two studies (Refs. 26 and 166) concluded that siderosis or pneumoconiosis in welders is not associated with a reduction in lung function while those from two other studies (Refs. 84, 150, and 151) concluded that such an association exists.

Although there is some conflict among these reports, the results suggest that bronchitis and deficits in pulmonary function among nonsmoking welders is possible. Clearly, welders who smoke have an elevated incidence of both bronchitis and pulmonary deficits. Whether smoking and welding exposures have additive or synergistic effects is unknown. The study of shipyard welders by Lyngenbo et al. (Ref. 92) carefully controlled for exposure to cigarette smoke, asbestos, and other lung toxicants. That study demonstrated that welding alone can cause bronchitis, upper respiratory tract symptoms, and reduced lung function.

The effects of cigarette smoke on the development of these symptoms were additive rather than synergistic. The influence of the chemical constituents of welding fumes on the development of pulmonary impairments is still not established. Krasnyuk (Ref. 84) found that welders of high alloy steel were more likely to have extensive pneumoconiosis with more severe restrictive disturbances in pulmonary function than welders of low alloy steel, while Ulrich (Ref. 179) found no differences in pulmonary function between welders with high and low chromium exposures.

7.4 Acute Respiratory Tract Disease — Case Report. The case of a man who developed a sudden, work-related onset of acute pulmonary disease 5 months after the start of his employment as a full-time welder was described by Silberschmid (Ref. 156). At first, his symptoms (coughing, chills with fever, tightness in the chest, and shortness of breath) occurred only at work. The symptoms became persistent during the next 2 months and headaches, general weakness, and weight loss developed. Clinical examination showed swollen, red bronchial mucosa, irregular opacities in chest X-rays, and deficits in pulmonary function. Extreme inflammatory changes were found in biopsied lung tissue. Steroid treatment produced some immediate improvement, but exertional dyspnea, bronchial hyperreactivity, and recurrent pneumonias persisted for 7 years.

Before the start of his employment, the patient had only occasional experience welding and had no history of pulmonary disorders. At the factory where his disease developed, he worked with mild steel, using mostly SMAW with basic and rutile electrodes; occasionally GMAW of stainless steel or aluminum was performed. Inspection of the work place indicated that he had been working without properly functioning local ventilation. Fume concentrations were high (about 20 mg/m^3) and manganese and copper levels varied from 2 to 4 mg/m^3 and from 0.3 to 1.2 mg/m^3, respectively. Silberschmid commented that the development of chronic fibrosis following acute pulmonary reactions to welding fume components is uncommon.

7.5 Production Coatings. Emmerling et al. (Ref. 38) examined effects of welding primed metals in 35 persons. Metals primed with zinc chromate were routinely welded by ten welders, nine worked with metals primed with alkyd paints and six welded metals primed with polyvinylbutyryal coats. Controls were flame cutters, welders of unprimed metals, and persons without welding experience. Breathing zone concentrations of Zn, Cr, Ni, Fe, Cd, NOX, CO, formaldehyde, toluene, xylene, and ethylbenzene were all below the respective German MAKs.

There was no correlation between workplace exposure and concentrations of zinc, hippuric acid, methylhippuric acid, formic acid, and carboxyhemoglobin in blood and urine samples. No changes in pulmonary function occurred during the shift or during the work week which could be related to welding of coated metals. On the other hand, nonspecific bronchial hypersensitivity, as measured by methacholine provocation, was two times higher before, and three times higher after the work shift in welders of primed metals than in controls. The differences were not statistically significant, which was attributed to the small number of participants in the study.

8. Cancer

8.1 Epidemiologic Studies. Although numerous studies have considered the incidence of cancer in welders, the cancer risk associated with welding is still uncertain. An elevated cancer incidence has been reported at different sites in the body, including the lung, nasal sinus, bladder, kidney, and digestive tract. Of these, the most frequently reported tumor is cancer of the lung. However, the association between lung cancer and welding is not clearly established. It has been suggested that focusing epidemiologic studies on populations of welders most likely to be exposed to carcinogens (e.g., chromium or nickel exposures in stainless steel fumes) may yield more meaningful results.

In his recent review of published cancer epidemiology data, Peto concluded that the association between cancer and welding exposures has not been adequately investigated (Ref. 139). According to Peto, many epidemiologic studies of welders have not considered the duration of the welding experience and the types of welding processes used. The confounding effects of tobacco smoking and other occupational and nonoccupational exposures have often been insufficiently addressed. In addition, in many studies the numbers of cancer cases among welders were too small to provide an accurate estimate of the relative risk (RR).

Peto described conditions for an epidemiologic study that would provide answers to many questions now left open by previous work. This retrospective cohort study would include 1000 men, half of whom had already died. The subjects would have been exposed to welding fume, with no significant outside occupational exposures, for at least 5 years or, preferably, all their working lives. To ensure sufficient time for the development of tumors, the welding cohort would have begun their employment as welders at least 20 years before the start of the study. Information on the welding processes used would be carefully collected so that a subset of welders with specific exposures, such as chromium and nickel in stainless steel, could be readily identified and studied.

Several epidemiology studies published during 1986 and 1987 addressed the lung cancer incidence in welders. A population-based case-control study of the association between occupation and lung cancer was conducted in New Mexico by Lerchen et al. (Ref. 90). The study group included 506 patients, ages 25 to 84 years, with primary lung cancer, other than bronchioloalveolar carcinoma. The controls were 771 persons, frequency-matched to the cases for sex, ethnicity, and 10-year age category. Participants in the study group were identified by the New Mexico Tumor Registry and controls were obtained through the Medicare Program. The participants (or surrogates) were interviewed to obtain information on occupational exposures, smoking habits, and employment histories. Welders had a significantly elevated lung cancer risk, with an odds ratio (OR) of 3.2 ($p<0.1$). The increased lung cancer risk persisted after adjusting for cigarette smoking and also when those with shipyard experience were excluded to reduce the number of participants with asbestos exposure. The authors cautioned that the "welders in the study had worked in diverse industries that may have involved exposures other than welding that also increased the lung cancer risk."

The association between occupation and lung cancer risk in a case-control study in three municipal areas of New Jersey was examined by Schoenberg et al. (Ref. 152). The study included 763 white males with primary cancer of the trachea, bronchus, and lung; the controls were 900 white males selected from driver's license and death certificate files in the general population. Workers in the shipbuilding industry had a significantly elevated risk for lung cancer. Although the excess risk for all shipbuilding workers was primarily among those with reported exposure to asbestos, the risk was also high among welders and other workers with no reported asbestos exposure. The lung cancer risk was not elevated in welders who had no experience in the shipbuilding industry.

Sjogren et al. performed a morbidity and mortality study based on the 1960 Swedish National Census in which over 20 000 males, aged 20 to 69 years, stated their occupation as welders or gas cutters (Refs. 158 and 162). The 1960 census data were linked with the Register for Statistics of Causes of Death and the Cancer Register for 1961 to 1970. Data on tobacco use were taken from a survey of Swedish males in different occupational groups. Pulmonary tumors were more frequent than expected among welders and gas cutters. After correcting for smoking habits, the rate-ratio was 1.30 ($p=0.02$). In addition, the incidence of renal tumors was significantly elevated in welders (rate-ratio = 1.30; $p=0.04$) while that of malignant melanoma was decreased (rate-ratio = 0.58; $p=0.01$). The prevalence of pleural mesothelioma was also elevated, suggesting cross exposures to asbestos in the welding cohort. No difference was observed in the rate of nasal tumors between welders and controls. Sjogren commented that a weakness inherent in this type of study is that the general classification welder included those with only brief exposures (Ref. 158).

To examine whether the increased lung cancer rate was associated with exposure to chromium, Sjogren et al. (Ref. 161) compared the mortality and the lung cancer incidence in welders exposed to fumes containing high and low levels of hexavalent chromium. The study groups consisted of 234 welders who had worked primarily with stainless steel for at least 5 years and 208 railway track welders who were exposed to only low levels of chromium at some time between 1950 and 1965. Although asbestos exposures could not be totally eliminated, welders selected for the study had not worked in areas where asbestos dusts were generated. Five welders with high chromium exposure and one with low chromium exposure died from lung cancer. These rates did not differ significantly from those of the general population, but the difference between the two groups of welders was statistically significant.

A case-control study of the association between lung cancer and occupation was conducted in two industrial areas of Norway by Kjuus et al. (Ref. 83). The study included 176 lung cancer cases and an equal number of referents without lung cancer or obstructive lung disease who were identified through hospital records. The data were adjusted for cigarette smoking. An increased lung cancer risk was seen in welders (28 cases of lung cancer; RR = 1.9). The risk was higher for a subset of 16 stainless steel welders (RR = 3.3).

Corey et al. (Ref. 25) published a brief report of a study in which the number of deaths from cancer in 3000 welders was compared with death rates in a cohort of 6500 plumbers and with the Ontario male population. The lung cancer rate of welders did not differ significantly from that of the two control populations.

Pearce and Howard (Ref. 136) studied the association between occupation and cancer mortality in New Zealand. Demographic data and statistics on tobacco use stratified according to social class were obtained from the 1976 New Zealand Census; cancer mortality data were gathered from the National Health Statistics Center's records of deaths registered in New Zealand between 1974 and 1978. Among the male population, 5356 cancer deaths occurred. Only tumor sites with a relative risk greater than 1.5 were considered to be excessive and were cited in the study. Stomach cancer mortality (12 observed deaths in welders; RR = 1.95) was elevated in welders, while mortality from cancers of the colon, leukemia, bladder, liver, lung, and buccal cavity in welders did not differ from that of the general population.

Puntoni et al. conducted a mortality study of 2702 dockyard workers in Genoa who died between 1960 and 1979 (Ref. 141). Of these workers, 147 were gas welders and 141 were electric arc welders. Causes of death were compared with those of the general population of Genoa. A significant increase in relative risk was found for total deaths, for digestive tract cancer (7 cases observed versus 3 cases expected; RR = 2.33), and for cardiovascular disease among gas welders. No cancer excesses were seen among electrical arc welders.

Some recent studies of electrical and electronics workers have indicated that exposure to extremely low frequency electromagnetic radiation (ELF) may be associated with an excess leukemia risk. Welders are exposed to ELF magnetic flux densities that are from 2 to 200 times those generally experienced in the household or by persons in other electrical trades. To better assess the risk of leukemia in welders, R. M. Stern examined pooled data from 15 published epidemiology studies that included information on the incidence of leukemia in welders (Ref. 170). The studies included a total of 146 cases of observed leukemia versus 159.46 expected, with a risk ratio of 0.92. For acute leukemia, 40 cases were observed versus 43.39 expected, with a risk ratio 0.92. Thus, Stern's calculations indicated that welders do not have an increased risk for leukemia. Stern concluded that the absence of an increased leukemia risk among welders does not support the hypothesis that the observed excess risk of leukemia among workers in the electrical trades is due to their ELF exposure.

A fivefold excess leukemia rate among workers at the Portsmouth (New Hampshire) Naval Shipyard was reported in 1978 (Ref. 105). At that time, the excess risk was thought to be due to radiation exposure during manufacture and maintenance work on nuclear submarines. As a follow-up to this report, the National Institute for Occupational Safety and Health conducted a matched case-control study of the leukemia incidence among the 24 545 onshore workers in all job categories employed between January 1952 and August 1977 at the Portsmouth shipyard (Ref. 169). Included in the study were 53 cases of deaths from leukemia and 212 controls. The study sought to ascertain whether the leukemia deaths were associated with occupational exposure to either ionizing radiation or to organic solvents. Radiation dose histories and detailed work histories by job and shop were evaluated for each subject.

Contrary to expectations, there were no significant associations between either ionizing radiation or presumed solvent exposure and myelogenous or lymphatic leukemia. However, an elevated incidence of leukemia was found among electricians and welders. For electricians, the Mantel Haenszel odds ratio was significantly elevated for all leukemias (ORMH = 3.00), and particularly for lymphatic leukemia, (ORMH = 6.00) while for welders, the odds ratio was significantly elevated for myeloid leukemia (ORMH = 3.83), but not for lymphatic leukemia. (Ref. 169)

As part of a cooperative Scandinavian investigation, Hansen et al. (Ref. 52) conducted a case-control study of the incidence of cancer of the nasal cavity and paranasal sinus. The 49 patients with primary nasal tumors were identified through the records of Danish Cancer Centers. Patients with colorectal cancer served as controls and were identified through the Cancer Registries in Denmark, Sweden, and Finland. Occupational histories were obtained by interview for each subject and covered the period between the start of employment and 10 years before the diagnosis of disease. Because of the long latent period of most cancers, exposures incurred during the 10-year period immediately preceding disease development were considered to be unrelated to the etiology of the cancer. Exposure to welding, cutting, and soldering fumes was found to be associated with an excess of nasal cancer. The data were not adjusted for asbestos exposure, which, according to the authors, may be a contributing factor to the development of both nasal and colorectal cancer.

Using both case-control and retrospective cohort approaches, Norell et al. (Ref. 109) examined the relationship between occupational factors and pancreatic cancer in Sweden. The case-control study used all newly diagnosed cancers of the exocrine pancreas in patients between 40 and 79 years old at three hospitals. Age- and sex-matched controls were patients with inguinal hernias and subjects chosen from the general population. Information on occupational history and exposures was obtained for each participant. The retrospective cohort mortality study was based on data from the Swedish Cancer Registry for 1961 to 1979 and from the 1960 census. The observed number of tumors in various job categories was compared with numbers expected from the cumulative incidence for all employed males.

An excess risk for pancreatic cancer among welders was suggested by the case-control study. Pancreatic cancer was observed in 13 patients with welding experience. The relative risks calculated from hospital controls and population controls were 1.7 and 2.0, respectively. In the retrospective cohort study, welders were not considered separately, but rather were included in the occupational categories "iron and steel works" and "iron and steel manufacture." Men in these categories showed no excess risk for pancreatic cancer. The standardized mortality ratios (SMR) for pancreatic cancer were 1.1 and 0.9 for "iron and steel works" and "iron and steel manufacture," respectively.

McLaughlin et al. conducted population-based assessments of the occupational risks for brain tumors (Ref. 97) and meningiomas of the central nervous system (Ref. 98). Data for these studies were obtained from the Swedish Cancer/Environment Registry. The risk of intracranial gliomas for various occupations, industrial groups, and regions was estimated with a 19-year cumulative standardized incidence ratio (SIR). The SIR is the ratio of the number of tumors observed to those expected in a particular employment category. Occupational risks were evaluated for 3394 cases of intracranial glioma (the most frequently encountered brain tumor in adults), and 1092 cases of meningioma among employed Swedish men for the years 1961 to 1979. Welders and metal cutters had a significantly elevated risk of developing brain cancer ($p < 0.05$) but not central nervous system meningiomas.

In contrast to the study of McLaughlin et al. (Ref. 97), studies by Thomas et al. (Ref. 175) and Olin et al. (Ref. 123) did not find an elevated risk of brain cancer in welders. Thomas et al. (Ref. 175) conducted a death certificate-based case-control study of workers in the electrical and electronics industry. This study included 435 white males from northern New Jersey, southern Louisiana, and Philadelphia who died from primary brain tumors between 1979 and 1981, and 386 controls who died from other causes. The case-control study of Olin et al. (Ref. 123) compared 78 astrocytoma patients, including five welders from two Swedish hospitals, with 197 clinical and 92 population controls. Risk estimates for astrocytomas in welders of 0.6 and 0.2 were based on hospital controls and population controls, respectively.

In summary, an elevated lung cancer risk in welders was found in four of the studies described above. In the study conducted by Sjogren et al., the increased lung cancer incidence among welders was independent of tobacco use, but asbestos exposure could not be ruled out (Refs. 158, 161, and 162). The studies by Sjogren et al. (Ref. 161) and Kjuus et al. (Ref. 83) suggested that stainless steel welders have a higher risk of lung cancer than mild steel welders. However, the number of stainless steel welders with lung cancer was too small to allow definitive conclusions. An elevated risk for lung cancer was also observed by Schoenberg et al. (Ref. 152) in a mortality study conducted in metropolitan New Jersey. However, in this case, only shipyard welders had an elevated lung cancer risk, even though the study controlled for asbestos exposure. In contrast, Lerchen (Ref. 90) observed an elevated lung cancer risk among welders even when shipyard welders were excluded. Neither Pearce (Ref. 136) nor Puntoni (Ref. 141) observed an elevated lung cancer risk among welders in the general population of New Zealand or in Italian dockyard workers, respectively. However, Pearce observed an elevated risk for stomach cancer while in Puntoni's study gas welders, but not arc welders, had an elevated risk of digestive tract cancer.

An elevated risk for leukemia was noted in a shipyard study conducted by F. B. Stern (Ref. 169). In contrast, R. M. Stern found no association between welding exposures and leukemia in his analysis of 15 epidemiologic studies which considered the incidence of leukemia in welders (Ref. 170). Similarly, an elevated risk for brain tumors was found by McLaughlin in a population-based study in Sweden (Ref. 97), but no brain tumors were identified in a death-certificate study by Thomas conducted in the U. S. (Ref. 175), or in a case-control study conducted in Sweden by Olin et al. (Ref. 123). One study found an elevated risk for renal tumors in welders (Ref. 161). The discrepancies between results of studies that appeared during the current report period do not help clarify the actual cancer risk of welders, especially at sites other than the respiratory tract.

8.2 Metal Carcinogens. A major argument supporting the probability of an increased cancer risk in stainless steel welders is the presence of nickel and hexavalent chromium in fumes from some welding processes. The carcinogenicity of chromium and other metals was recently reviewed by Norseth (Refs. 110 and 111). According to Norseth, the epidemiologic evidence for an increased risk of cancer in workers exposed to arsenic, chromium, and nickel in some industries (other than welding) is extensive, whereas that for exposure to beryllium and cadmium is more limited. The epidemiologic data receive some support from tests in animals and in vitro systems, but the mechanism of action of metal carcinogens remains unknown and is an important area for further research.

Factors such as solubility, particle size, crystalline structure, chemical species, and surface charge may be important in determining the biological activity of metals. Generally, only slightly soluble compounds of nickel (e.g., sulfides and oxides) or of chromium are thought to cause cancer. Norseth (Ref. 110) suggested that the solubility of these compounds may be important because compounds with limited solubility would be more readily incorporated into cells by phagocytosis than would the more soluble compounds. Once inside the cells, some particulate material could gradually become solubilized, bringing about a continuous interaction of the metal with the cells and possibly a chronic interference with normal cellular processes.

Norseth cited limited evidence implicating trivalent chromium in the carcinogenic process and suggested that intracellular reduction of hexavalent to trivalent chromium may be important to the mechanism of chromium carcinogenicity. He further postulated that the activity of a short-lived intermediate, perhaps in the pentavalent state, may be responsible for the carcinogenic or transforming activity of chromium.

According to Norseth, the experimental evidence indicates that all nickel and hexavalent chromium compounds should be considered to be carcinogenic and should be treated as such by regulatory agencies (Ref. 111). This view is disputed by Mastromatteo (Ref. 95) who noted that, in the recent health effects document for nickel (Ref. 181), the U. S. Environmental Protection Agency concluded that only the following three forms of nickel should be classified as known or probable human carcinogens: nickel refinery dust (from pyrometallurgical sulfide nickel matte refineries), nickel subsulfide, and nickel carbonyl. Mastromatteo stressed that a cancer risk is associated with specific nickel species and indicated that air sampling methods should determine not only the total amount of nickel, but also the species present.

8.3 Screening for Cancer. Because respiratory tract cancer is generally diagnosed late in development, the death rate is generally high for this disease. Thus, reliable and noninvasive screening tests for detection of early stages of lung cancer have been sought for use in industries in which the cancer rate may be elevated. Two methods, chest X-rays and sputum cytology, are currently available for screening large groups of workers for lung cancer. Chest X-rays are less useful for detecting early tumors than sputum cytology which can distinguish between precancerous and cancerous lesions. According to Kilburn (Ref. 80), sputum exams should be performed every four months to be effective. Unfortunately, sputum samples do not generally contain cells from the more distal parts of the respiratory tract and so some lung tumors can be missed with this technique.

Methods are currently available for screening workers at risk for development of nasal cancer by performing biopsy of the nasal mucosa. However, obtaining tissue samples can be painful, and preneoplastic lesions may be missed since they tend to be present as small foci. To overcome these problems, Reith developed a cytologic screening method in which a more representative population of cells is less painfully obtained by brushing the surface of the middle nasal turbinate. In a study of nickel refinery workers with known nasal dysplasia, Reith showed that more preneoplastic nasal lesions could be detected by cytologic screening than by biopsy (Ref. 144).

In an effort to identify a substance in sputum that could serve as a marker for cancer risk, Ebeid studied the relationship between thiocyanate levels in sputum and welding or smoking, or both. Sputum samples were obtained from 60 persons employed in a metal forming factory. Thirty-two subjects were welders and the remainder were office workers who served as controls. Ten welders and sixteen office workers were nonsmokers. Mean sputum thiocyanate concentrations in the nonsmoking welders and office workers were 27.77 and 32.92 ug/ml, respectively, while concentrations in the sputum of smoking welders and office workers were 59.27 and 110.42 ug/ml, respectively. Ebeid concluded that exposure to welding fumes may decrease the thiocyanate content of sputum, while smokers have elevated levels relative to nonsmokers. He suggested that determining the thiocyanate content of sputum may be of value as an index of certain exposures in the work environment. However, it would appear that, in the absence of further information, thiocyanate levels do not promise much utility for estimation of disease risk in welders (Ref. 37).

9. Metal Fume Fever

Metal fume fever is a flu-like disease caused by inhalation of oxides of various metals including zinc, copper, aluminum, antimony, iron, manganese, nickel, and cadmium. The symptoms, which include fever, chills, general malaise, joint pains, cough, sore throat, chest tightness, and fatigue, usually appear 4 to 12 hours following exposure and last from 1 to 2 days. A short-lived tolerance to metal fume fever can develop, which is lost during a short absence from exposure. The name "Monday fever" has been conferred because symptoms may recur on return to work after a weekend or holiday. McMillan suggested that the tolerance is a practical, rather than a physiological phenomenon resulting from the exercise of less caution when welding after a period of absence from work. Conversely, after a bout of the disease, the welder may more carefully avoid fume exposure or his supervisor may be more likely to place him in a better ventilated area (Ref. 100).

The pathogenesis of metal fume fever is poorly understood. It is frequently attributed to an immune reaction to metals, resulting in inflammation of respiratory tract tissue and release of histamine or histamine-like substances. Vogelmeier et al. (Ref. 186) observed changes in pulmonary function and marked pulmonary inflammation in a 26 year old male welder who had recurring symptoms of metal fume fever after welding zinc-coated materials but not other metals. When the welder was asymptomatic, clinical findings, chest X-ray, and spirometric data were all within normal limits, but inhalation challenge with methacholine revealed a low grade nonspecific bronchial hyperreactivity.

Two tests were conducted in which the subject welded a zinc-coated tube for 1 hour under controlled conditions. Airway resistance increased, while inspiratory vital capacity, single-breath transfer factor, and arterial oxygen partial pressure were sharply reduced immediately after welding. Chills, fever, and an increased peripheral blood leukocyte count occurred 2 to 6 hours later.

A second test, performed after an exposure-free 6-month period, yielded similar results. Bronchoalveolar lavage was performed 24 hours after the second controlled welding exposure. The total number of cells was ten times greater than normal, and the number of polymorphonuclear leukocytes was markedly increased in lavage fluid. The cell population in bronchoalveolar lavage fluid was normal after a 7-week exposure-free period. Thus, exposure to zinc fumes produced a pronounced pulmonary reaction, with a severe acute inflammation of peripheral bronchoalveolar structures. The investigators remarked that metal fume fever has not, thus far, been known to cause pulmonary fibrosis; they questioned whether the marked increase of leukocytes can really be harmless to pulmonary tissues. This question is similar to that raised by Mylius and Gullvag (see above) who observed a marked excess of alveolar macrophages in welders who smoke (Ref. 104).

A case of metal fume fever with a severe concomitant allergic reaction to zinc fumes was reported by Farrell (Ref. 40). After welding galvanized metal, a 34 year old male patient experienced classic symptoms of metal fume fever associated with an immediate anaphylactoid reaction and a delayed reaction consisting of angioedema (swelling of the face, lips, and throat) and urticaria (an allergic skin reaction with hives and intense itching). The symptoms recurred after further exposures to zinc fumes even though a protective face mask was used. To confirm that the anaphylactic reaction was associated with exposure to zinc fumes, the subject welded galvanized steel under medical supervision. Four hours later he developed symptoms of metal fume fever. As on previous occasions, angioedema and urticaria developed 12 hours after exposure. These clinical symptoms were suggestive of an immediate and late immunoglobulin-E-mediated immunological reaction to zinc fumes. Although metal fume fever is thought by some investigators to have an immunological basis, Farrell does not believe that a direct association between immediate and late anaphylactoid reactions and metal fume fever can be fully substantiated by this case.

Shusterman reported the case of a man who experienced signs of metal fume fever and injury to myocardial and skeletal muscle after welding galvanized metal for 4 successive days. Welding was performed outdoors with no respiratory protection. On 2 of the 4 days, he ran for 2 to 3 miles after the completion of work. After the fourth day, he had chest pains and was hospitalized for 10 days. Clinical tests indicated that there might have been injury to skeletal muscle as well as the heart. The authors suggested the possibility of a causal relationship between inhalation of zinc oxide fumes and muscle injury (Ref. 155). However, this would be difficult to substantiate on the basis of the data presented in this case.

10. Effects on the Ear and Hearing

The case of a welder who became permanently deaf in one ear as a result of a metal spark burning the middle and inner ear was reported by Stage and Vinding (Ref. 167). The incident occurred while he was lying on his side welding underneath a car. He was wearing a welding mask which afforded no protection to the ears.

11. Effects on the Eye and Vision

Exposure to electromagnetic radiation in the IR, visible, or UV range generated by arc welding can injure the eye. Shortwave UV light (270 to 290 nanometers) is absorbed by the outer layers of the eye; this can result in keratoconjunctivitis (arc eye, welder's flash, photokeratitis, or photophthalmia), and pterygia (membranous growths which extend across the outer eye from the conjunctiva to the cornea). Near UV light, in the range 295 to 400 nm, penetrates more deeply into the eye where it can damage the corneal epithelium and possibly the lens. Infrared radiation of longer wavelengths may cause thermal damage to the cornea and aqueous humor and has also been associated with the development of lenticular cataracts. Brief and intense exposure to blue light in the visible range (approximately 400-500 nm) or to near IR radiation may produce photochemical lesions on the retina. The potential effects of welding and electromagnetic radiation on the eye were reviewed by McMillan (Refs. 69 and 99).

Keratoconjunctivitis, a marked inflammation of the cornea, is the eye condition most frequently encountered by welders. It is characterized by blurred vision, tearing, acute pain, and headache. Symptoms normally last up to 2 days and there are usually no sequelae. Most pharmaceutical preparations that provide relief of symptoms require prescriptions and close medical monitoring because they can produce permanent corneal damage and loss of visual acuity. Nonprescription drugs are available, but they are generally less effective and require frequent application. An investigation of the efficacy of nonprescription drugs was conducted by Wittgens (Ref. 194). Over-the-counter preparations for the treatment of keratoconjunctivitis were distributed to welders through occupational physicians and other medical professionals in the workplace. A survey of the welders who used the products indicated that an eyedrop preparation (Chibro-Uvelin), developed in France in 1953, was the most satisfactory of the products tested. It provided fast relief from pain and return of visual acuity. Unlike some other drugs tested, it was effective for a long time and did not require repeated application. According to Wittgens, the effects of this preparation have been thoroughly researched, and it has no known side effects.

Sharir and Ben-David observed severe and persistent symptoms of keratoconjunctivitis among four workers who performed plasma arc thermal spray coating (Ref. 153). Excessive tearing, foreign-body sensation, and dull pain persisted for several months. Two of the patients had mild to moderate signs of nonspecific irritation, while the other two patients had severe acute inflammation with pronounced swelling of the conjunctiva and eyelids. The disorders were attributed to exposure to the high levels of UV and visible radiation generated by plasma arc thermal spray coating.

While the retina is the part of the eye most sensitive to UV radiation, injury to the retina from exposure to the welding arc is rare and usually results from improper eye protection (Ref. 99). A case of a 22 year old man who developed retinal injury with substantial bilateral vision loss soon after welding without appropriate eye protection was reported by Uniat et al. (Ref. 181). Some visual loss and signs of retinal injury were still present 6 months after the incident. Cellini et al. described a similar case of photic maculopathy in a young worker who was said to have been wearing "protective eyeglasses" while welding (Ref. 20). Visual loss was noted soon after exposure to the welding arc; the progression of the disorder followed a favorable course with treatment.

An unusual case of retinal damage in a man with a lens implant was reported by Turut et al. (Ref. 179). The subject was a 36 year old welder in whom a cataractous lens had been replaced by a lens implant. Four years after the surgery, an incident in which he welded without wearing eye protection was followed immediately by loss of visual acuity. Examination revealed a small lesion, apparently caused by radiation from the arc, on the retinal fovea of the eye with the lens implant. There were no indications of damage or loss of vision in the eye with the normal lens. The investigators noted that the normal lens provides some protection to the retina by filtration of potentially harmful IR and visible radiation. Persons with lens implants are more sensitive to the effects of radiation and should use extreme caution when performing operations, such as welding, which generate intense electromagnetic radiation.

Retinitis pigmentosa is a hereditary condition involving progressive atrophy of the retina. The first symptom of the disease is night blindness (reduced ability to adapt to low intensity light). As the disease progresses, affected persons have increasing difficulty adjusting to bright light. Two welders with retinitis pigmentosa sine pigmento were described in the Yugoslavian literature by Krizmanic-Mrkoci et al. (Ref. 86). After two years of welding, one welder complained of vision problems and photophobia (abnormal sensitivity to light). The problems were greatest when he left the plant at the end of the work day. Adaptometry and electroretinography indicated that he, and a second welder with similar visual difficulties, had retinitis pigmentosa sine pigmento. This condition makes it difficult for welders to use protective eye wear during work and affects their ability to adapt to bright sunlight upon leaving the workplace. Krizmanic-Mrkoci recommended that peripheral vision tests, which detect retinitis pigmentosa, be included in regular screening exams given to all welders before employment.

The debate persists in the open literature on the potential hazards of wearing contact lenses during welding (Ref. 19). Contact lenses are generally considered to be inadvisable in situations or environments where there may be exposure to chemical fumes, vapors, splashes, intense heat, or high concentrations of particulates. Most authorities discount reports that radiation emitted by the welding arc can cause adhesion of the contact lens to the cornea. In fact, some experts believe that, with the use of appropriate eye safety equipment, there may be advantages to wearing contact lenses in hazardous industrial settings. Callier (Ref. 19) would encourage persons who regularly use contact lenses to wear them in hazardous industrial situations because they frequently see better with them than with eyeglasses.

Randolph and Zavon (Ref. 142) reviewed the literature on contact lenses and found that contact lens wearers exposed to high levels of atmospheric particulates do not have excess subjective or objective symptoms of eye damage. Contact lenses do not worsen the corneal damage resulting from exposure to chemicals and in some cases may provide protection for the eye. In fact, contact lenses may be beneficial because they allow a closer fitting of protective eye wear. The authors surveyed the corporate policies among 67 large American firms and found that 86% had some restrictions on the use of contact lenses. While most of the restrictions involved areas where irritant chemicals were present, 52% of the companies restricted contact lens use in welding operations. Analysis of the survey responses indicated that most corporate policies were based on perceived risk rather than on facts revealed by recent research.

12. Effects on the Nervous System

Chronic exposure to manganese can cause a progressive disease with neurological and psychological manifestations. Early symptoms include apathy, anorexia, spasm, and irritability. Signs of psychosis may develop with euphoria, impulsive behavior, confusion, and aggression. Advanced stages resemble Parkinson's disease, with muscle weakness, muscle rigidity, tremors, and impaired gait. Two cases of welders who had advanced stages of manganese poisoning with symptoms of Parkinsonism were reported by Rasmussen and Jepsen (Ref. 143). Both welders had performed SMAW in a boiler factory, one for 31 years and the other for 17 years. Hygienic conditions in the factory were poor and the authors presumed that exposure to manganese was high.

Rudell (Ref. 147) reported two cases of welders with neurological symptoms. The first was a 29 year old man who, after 8 years employment in SMAW of mild steel, became dizzy when welding. The second was a 43 year old man employed in GTAW of Nimonic alloys (75% nickel, 25% chromium) for 14 years. He reported dropping objects and had diffuse neurological symptoms in his left side. Physical examination revealed no other neurological disorders.

13. Effects on the Skin

A case of a welder with recurrent, severe facial dermatitis was described by Shehade et al. (Ref. 154). Allergic dermatitis was ruled out with skin patch testing, using a standard battery of allergens, preservatives, potential sensitizers, photoallergens, a welding antispatter formulation, and welding flux. The case was diagnosed as photodermatitis, caused by UV exposure during welding.

The affected welder performed spot and seam welding with CO_2-shielded GMAW of mild steel. During spot welding, he seldom wore a protective face mask and, instead, turned his face to the left to protect his eyes from the arc. His UV exposure was measured with polysulfone badges; 20 badges were used over a 3-week period. In all cases, the measured exposure was equal to or greater than the maximum permissible exposure (MPE) for an 8-hour day, established by the ACGIH in 1976. In the worst case, the exposure was 128 times the MPE. When the badge was placed underneath the protective face mask, the MPE was still exceeded by 4 to 9 times.

14. Sensitivity to Fume Components

Allergic contact dermatitis in response to chromate salts is well documented among welders, but anaphylactoid reactions are rare. A case of a welder who developed severe systemic reactions after exposure to chromates was reported by Moller et al. (Ref. 101). The welder had 10 years' exposure to vapors from chromium trioxide baths and stainless steel welding fumes. Symptoms first appeared after several hours exposure to chromium trioxide mist during an accidental discharge from an adjacent chrome-plating plant. At this time, he developed swelling and a diffuse pruritic papular rash diagnosed as urticaria. This reaction to chromium occurred twice more over the next 7 months. With time, he also experienced difficulty breathing and chest tightness. Challenge with 29 ug/m^3 sodium chromate aerosol induced a delayed anaphylactoid response with urticaria, angioedema, severe bronchospasm, and a threefold increase in plasma histamine levels. Skin tests and in vitro immunological tests indicated that this was probably not a classic IgE-mediated allergic response, but rather may have been a cell-mediated response involving sensitized lymphocytes.

Nordman et al. (Ref. 108) examined 58 welders who had asthmatic reactions when welding stainless steel. Of these welders, 13 had significant reduction in pulmonary function following welding of stainless steel but not mild steel. Seven of the 13 welders responded positively to the histamine provocation test prior to welding.

Exposure to high levels of soluble beryllium compounds can cause acute berylliosis characterized by inflammation of the nasopharynx, trachea, and bronchi with possible development of pulmonary edema. Repeated exposure to low levels of slightly soluble beryllium compounds may induce chronic berylliosis. This condition is often thought to have an immunologic basis and is typified by a distinctive granulomatous process in the lung which may not become apparent until months to years after exposure ceases. Because of similarities in pathologic processes in the lung, chronic beryllium disease is readily mistaken for sarcoidosis. Cullen et al. examined five precious metal refinery workers with symptoms of chronic berylliosis and demonstrated that the etiology of the disease could be confirmed by an in vitro immunologic test (Refs. 27 and 28). In this test, alveolar lymphocytes from persons with berylliosis, but not from normal controls, undergo division when exposed to beryllium. The investigators concluded that their results provide further evidence for an immunologic basis for chronic beryllium disease.

15. Effects on the Musculoskeletal System

Awkward work postures of the trunk and shoulder may cause fatigue and musculoskeletal disorders. For example, prolonged elevation of the arms (shoulder flexion or abduction) can produce extreme muscle fatigue and, in some cases, acute tendonitis. Evaluation of postural stress in the workplace is difficult because a continuous record of the time spent in various postures during a work operation must be obtained and analyzed. To identify the causes of postural stress, Keyserling developed a computer-aided system to analyze working postures of the trunk and shoulders on repetitive assembly line jobs (Ref. 78). With this system, an analyst watches a continuous videotape record of work operation and enters a symbol into the computer for each assumed posture. The time duration of each posture is automatically measured by the computer which generates a posture profile for each joint (trunk, right shoulder, left shoulder). The posture profile is then used to determine whether there is an excessive frequency of posture change or whether non-neutral postures must be maintained for excessive time periods. The system was tested at a manual spot welding operation in an automotive assembly plant. Based on the results of the computer-aided analysis, work stations were redesigned to eliminate awkward postures and to incorporate several new safety features.

A system for analyzing working posture, developed by the Finnish steel industry, was tested by Heinsalmi (Ref. 55). A positive relationship was found between poor working postures and sick leave due to musculoskeletal disorders, especially lower back pain. Sick leave due to neck and shoulder disorders also correlated with poor arm postures. A prototype wooden manikin was used to study work postures in welding operations. A welding unit, developed on the basis of that study, improved welding posture and enabled operational time savings.

Bjelle et al. (Ref. 15) examined the influence of ergonomic factors on shoulder and neck complaints in 26 Swedish industrial workers who performed assembly, pressing, or welding during the manufacture of truck cabs. Arm movements during work were filmed, and shoulder strain was estimated by measuring the duration and frequency of shoulder abduction or forward flexion. Over a 12-month period, two thirds of the workers complained of musculoskeletal injuries; nine of these were shoulder and neck complaints. Tests of muscle strength showed that the strength of the shoulder muscles, but not elbow flexion or hand grip measurements, was significantly lower in workers with shoulder-neck complaints. The authors concluded that since shoulder muscles, but not elbow or hand muscles, were weaker in workers with shoulder-neck complaints, shoulder muscle weakness may be a secondary phenomenon rather than a reflection of a predisposition to shoulder and neck disorders.

16. Effects on the Urogenital Tract

Chronic exposure to low concentrations of cadmium fumes can cause kidney damage, which is first manifested by urinary excretion of low molecular weight serum proteins, such as beta-2-microglobulin and retinol-binding protein. The proteinuria is generally thought to result from damage to the kidney tubules. Whether or not cadmium also affects glomerular function remains unresolved.

Smith et al. (Ref. 164) examined the cadmium body burden and renal function in 53 solderers from five firms, who had up to 25 years of uncontrolled exposure to fumes from cadmium-containing alloys. Elevated cadmium levels in the kidney and liver were found in 22 of 25 subjects, most of whom had more than 5 years' exposure to cadmium solders. Cadmium concentrations were elevated in blood and urine in all 32 workers with more than 5 years' exposure, and in 11 of 21 workers with less than 5 years' exposure. Low molecular weight proteins, typical of tubular dysfunction, were present in urine from 20 of 51 subjects tested, while albumin, characteristic of glomerular dysfunction, was found in only 3 of 51 subjects. No significant changes in urinary cadmium or beta-2-microglobulin were found in 19 workers followed for 2 years after cadmium-free soldering rods were substituted for cadmium-containing soldering rods. The latter observations suggested that mild tubular proteinuria does not progress to more severe kidney dysfunction in the absence of further accumulation of cadmium within the body.

Verschoor et al. (Ref. 185) compared urinary protein and blood and urinary cadmium concentrations in 26 cadmium-exposed workers and eight nonexposed controls. Nineteen of the cases were exposed to cadmium in plastic pigments and the remainder were welders in a radiator shop. The mean values of urinary retinol-binding protein and urinary and serum beta-2-microglobulin were significantly higher in the cadmium-exposed welders than in controls, while total urinary protein and urinary albumin did not differ. Thus, signs of tubular dysfunction, but not glomerular dysfunction, were present in this group of cadmium-exposed welders.

Prompted by observations of five patients with glomerulonephritis or interstitial nephritis at the University Hospital in Umea, Hagberg et al. conducted a negative case-referent study to determine if there is an association between chronic renal disease and welding. The study group consisted of 143 living males who had undergone kidney biopsies as a result of suspected or clinically established glomerulonephritis or obscure renal disease. Occupational and vocational exposures were compared with age- and residence-matched controls chosen from the general population. No relationship between welding and the incidence of severe chronic kidney disease was found. In addition, there was no increase in the risk of death from nephritis among welders. (Ref. 51) In this study, no distinction was made between welders who had cadmium exposures and those who did not.

Similarly, no association between renal impairments and stainless steel welding was found in a cross-sectional field study performed by Zschiesche et al. (Ref. 196). Kidney function, assessed by measurement of total protein and specific enzymes in urine, was compared in 118 stainless steel welders and an equal number of age-matched controls. Breathing zone measurements during welding revealed substantial exposures to nickel and chromium during welding. The concentrations of nickel and chromium were significantly higher in urine from welders than from controls. However, the results of kidney function tests did not differ between welders and controls or among welders using different welding processes. The authors noted that, although urinary chromium and nickel levels were elevated in the welders, they were not excessively high. They recommended that additional studies be conducted to examine these parameters in welders with greater exposures to nickel and chromium.

17. Effects on the Teeth and Oral Cavity

Underwater welders frequently sense a metallic taste while welding and have a higher than normal replacement rate of dental amalgam (Ref. 133). To further examine the association between changes in dental amalgam and underwater welding, Ortendal and Holland measured the intraoral electrical activity while divers were welding underwater (Refs. 127, 130, and 131). They noted that during underwater welding or cutting, the electrical activity in the mouth is normally too low to cause a metallic taste or corrosion of dental amalgam (Ref. 130). During three test dives, leaks in the dry suit allowed a small amount of water to enter the gloves which resulted in an increase in the intraoral electrical activity. In one such instance, a diver experienced a metallic taste. This could not be explained by deterioration of dental amalgam since the electrical activity was theoretically too low to damage the amalgam. Further studies with divers showed that the metallic taste sensation was unlikely to be a result of magnetic field effects (Refs. 127, 128, and 129).

An in vitro system was designed in which exposure of amalgams to magnetic fields simulated the environment of divers when using a mean cutting current of 650A DC (Refs. 127 and 129). Studies with this system demonstrated that a slight roughening of amalgam surfaces could result from magnetic field exposures. However, chemical changes in the amalgam were insufficient to account for a metallic taste sensation. Thus, neither in vivo nor in vitro studies supported the view that this subjective symptom is related to magnetic field effects. Based on an earlier report (Ref. 107), the investigators postulated that the metallic taste resulted from electrical stimulation of the taste buds.

18. Effects on the Voice

Voice quality was compared in eight shipyard welders and eight office workers; none of the participants smoked. A questionnaire indicated that voice and throat problems were more frequent among welders than office workers; tests indicated that the welders' voices were more hyperfunctional and unstable. The investigators concluded that voice problems among welders are due to high noise levels which cause a strain on the larynx during vocalization. The welders thought their voice problems to be due to fume exposures (Ref. 112).

19. Biological Monitoring

19.1 Aluminum. A study of aluminum excretion among 23 GMA and GTA welders by Sjogren et al. revealed that aluminum levels in urine are dependent on the most recent daily exposures, as well as on the total duration of exposure (Refs. 159 and 160). Aluminum levels, measured in the breathing zone of 16 aluminum welders at the start of the study indicated that most exposures were greater than 1 mg/m^3. Aluminum concentrations were determined in urine from 16 welders before and after a 16 to 37 day exposure-free vacation period. Biological half-times for urinary aluminum were calculated from samples collected from eight of the welders in the middle of the exposure-free period. The half-time for welders with less than 1 year exposure was about 9 days, whereas it was 6 months or more in welders who had been exposed to aluminum fumes for more than 10 years. Half-times for aluminum calculated from urine collected at the end of the exposure-free period were 27 days for welders with less than 1 year's experience and more than 2 years for the more experienced welders.

The investigators remarked that, even though there was a wide variation in urinary aluminum levels among individuals, the large difference in urinary half-times between more and less experienced welders indicated that aluminum is probably stored in two or more compartments in the body, with widely different clearance rates. They speculated that the compartment with the slow clearance rate probably includes the lungs and the bones.

19.2 Manganese. Zschiesche et al. determined levels of manganese in the breathing zone and body fluids of eight welders from two factories who worked with filler materials containing about 0.8% manganese (Ref. 197). SMAW was the predominant welding process, but GMAW was also used. Manganese was measured in air samples collected for 2 hours each day during the work shift for a period of 1 week and in urine collected 3 times daily (before, during, and after the work shift) for 1 full week and the following Monday. Blood samples were collected before and after each work shift during the same time period.

Median breathing zone concentrations were 11.5 mg/m^3 for total fume and 257.7 ug/m^3 for manganese. Even though fume levels tended to be high, only one sample had a manganese concentration in excess of the German MAK of 5 mg/m^3. Manganese levels in blood and urine tended to be low and, for the most part, did not vary substantially from values obtained for the general population. There was no correlation between external exposures and internal manganese levels, or with the time of day or the day of the week that the biological samples were collected. It was concluded that, under the working conditions at the two plants, biological monitoring of manganese is not necessary.

19.3 Chromium. Because hexavalent chromium passes more readily through cell membranes, trivalent chromium tends to be transported in the blood by plasma while hexavalent chromium is transported mainly in erythrocytes (Ref. 3). Two research groups examined whether this difference may be used in biological monitoring to distinguish between exposures to trivalent and hexavalent chromium. Angerer et al. (Ref. 4) determined chromium levels in erythrocytes, plasma and urine obtained from 103 male welders who performed SMAW or GMAW, or both, of stainless steel. Median breathing zone concentrations of chromium were 4 ug/m^3 for SMAW and 10 ug/m^3 for GMAW.

The median chromium concentrations in erythrocytes, plasma, and urine of all welders were less than 0.60, 9.0 and 32.5 ug/l, respectively. The concentrations of chromium in erythrocytes, plasma, and urine for individual welding processes are shown in Table 3. In all cases, levels were higher in welders performing GMAW, or a mixture of GMAW and SMAW, than in those performing SMAW. Urinary chromium levels among the welders ranged from 5.40 to 229.4 ug/l, which is 5 to 200 times greater than average levels in the general population.

The correlation between chromium concentrations in erythrocytes, plasma, and urine was highly significant ($p<0.0001$). Chromium levels in erythrocytes, plasma, and urine of 0.6, 10 and 40 ug/l, respectively, corresponded to an external chromium trioxide exposure of 100 ug/m^3. Erythrocyte chromium concentrations were close to the limit of detection and were thus too low to provide accurate estimates of exposure below airborne concentrations of 100 ug/m^3 (the German "technical guideline concentration" for chromium trioxide). It was concluded that plasma or urine levels are more suitable for monitoring low level chromium exposures. At higher exposures, the erythrocyte concentrations may be useful as a measure of hexavalent chromium exposures; simultaneous determination of chromium levels in plasma and erythrocytes may give a useful picture of both overall exposure to chromium and exposure to hexavalent chromium.

Conflicting conclusions were drawn from a similar, but smaller, study conducted by Gustavsson and Welinder (Ref. 50). Blood plasma, erythrocyte, and urinary chromium concentrations were measured in samples collected from four stainless steel SMA welders on a Thursday morning before work, on Friday before and after work, on the following Tuesday, and after 31 days of vacation. A decrease in chromium levels was seen in all fluids during the vacation period and there was a significant correlation between chromium levels in the various fluids at each time point. Because there were no obvious differences in clearance rates of erythrocyte and plasma chromium, it was concluded that determination of blood or erythrocyte chromium levels would not add important information to that gained from biological monitoring with urinary chromium concentrations.

19.4 Analytical Methods. Methods for determining low concentrations of metals in biological fluids for use in biological monitoring were described by three groups of investigators (Refs. 22, 54, and 79). Heinrich-Ramm and Angerer described a method for determining concentrations of nickel and chromium in urine, plasma, and erythrocytes (Ref. 54). Christensen and Kirchoff (Ref. 22) developed a method using electrothermal atomic absorption spectroscopy (AAS) for analysis of chromium, nickel, and cadmium in blood. Kiilunen et al. (Ref. 79) developed a method using AAS for determining low concentrations of nickel and chromium in urine. Using this method, they found a mean nickel concentration of 70 nmol/l in urine from 299 Finnish persons who had no known occupational exposures to nickel. A mean value of 1.5 nmol/l (0.078 ug/l) was determined for urinary chromium concentrations using a study population of 155 persons.

Section Three Investigations in Animals and Cell Cultures

20. Animal Studies

20.1 Inflammation and Fibrosis. The development of practical methods for evaluating the fibrogenicity of fumes and gases from welding operations is important for the assessment of risk associated with different electrodes and welding methods. Hicks et al. used the inflammatory response of rat and guinea pig muscle tissue (Ref. 62) to compare the irritant and fibrogenic properties of several different welding fumes. Intramuscular injection of particles from SMAW of stainless steel caused severe local inflammation and fibrosis and was cytotoxic to muscle fibers. Fumes from SMAW of mild steel caused similar, but less severe effects. Fumes from GMAW of stainless steel caused only mild irritation and fibrosis and were not cytotoxic.

The effects on rats of single 6-hour inhalation exposures to fumes from six welding processes (Table 4) were investigated by the American Welding Society (Ref. 23). The major histopathologic changes resulting from fume exposure were seen in the respiratory tract. Fumes from electrode E308-16 produced the most severe effects, causing pulmonary edema and pneumonitis. Electrodes E7018, E6010, and E70T-1 caused upper respiratory tract irritation while the remaining two produced no apparent changes in the respiratory tract (Table 4). Accompanying these effects were changes in the number of macrophages that could be retrieved from the lungs by lavage. Decreased phagocytic activity was observed in macrophages from all test animals except those treated with fumes from E5356. Extracts of urine were not genotoxic in the Salmonella/Ames test nor were there any chromosome aberrations in bone marrow cells recovered from animals 20 hours after exposure.

Alveolar macrophages from untreated rats, were incubated with eight concentrations of welding fume particulates. Viability, measured in terms of the concentration of fumes that killed 50% of the cells (EC50), was lowest with fumes from electrode E308-16. On the basis of histologic assessment of the respiratory tract and in vitro assays of macrophage activity, the electrodes were classified into

Table 3
Chromium Concentrations in Erythrocytes, Plasma and Urine of Welders Using SMAW or GMAW, or both, of Stainless Steel

Welding Process	Chromium Concentration (ug/l) Erythrocytes X + SD (median)	Plasma X + SD (median)	Urine X + SD (median)
SMAW (n=39)	1.18 + 1.35 (<0.6)	9.0 + 7.81 (6.4)	35.14 + 40.11 (22.85)
GMAW (n=14)	1.44 + 1.72 (<0.6)	13.56 + 12.34 (9.6)	64.46 + 65.31 (37.60)
SMAW and GMAW (n=50)	2.51 + 5.82 (<0.6)	14.70 + 10.92 (12.25)	59.67 + 53.63 (40.0)
Total (n=130)	1.86 + 4.21 (<0.6)	12.39 + 10.33 (9.00)	51.19 + 57.75 (32.5)

Data from Angerer et al., Ref. 4

three categories: highly toxic (E308-16), moderately toxic (E7018, E6010, and E70T-1), and nontoxic (E70S-3 and E5356).

Wilmer et al. exposed rats for 14 days by inhalation to fumes generated by SMAW and GMAW of stainless and mild steel and by SMAW of cast iron using ten different electrodes (Ref. 188). Fumes from SMAW of stainless steel were the most toxic, causing severe dyspnea, some mortality, and reduction in body weight gain. GMAW fumes, especially those from mild steel, were less toxic than SMAW fumes. They induced slight signs of respiratory distress, increased lung weight and morphological changes in the lungs, including bronchiolar constriction, focal interstitial pneumonia and emphysema. Fumes from SMAW of mild steel were just slightly toxic, causing only a minimal increase in lung weight in males. The fumes from all electrodes except one used in SMAW of mild steel were cytotoxic to alveolar macrophages. None of the fumes caused SCE or structural chromosome aberrations in peripheral blood lymphocytes.

Gorban et al. exposed rats by intratracheal instillation to particles of magnesium, and three aluminum/magnesium alloys. All particle types caused histopathologic changes in the lungs. Alloys of magnesium and aluminum were more fibrogenic than pure magnesium. It was concluded that the crystal lattice of the alloy was important in determining its fibrogenic properties (Ref. 44).

20.2 Clearance. Kalliomaki et al. compared rates of excretion and pulmonary clearance of fumes from GMAW and SMAW of mild and stainless steel (Refs. 70, 71, and 73). Fume samples were administered to rats by intratracheal instillation. Concentrations of nickel, iron, and chromium were periodically determined in the lungs, urine and feces from serially sacrificed rats. About 10% of the administered fumes from all welding processes were recovered in the feces during the first three days after exposure. The ratio of elements in the feces was similar to that in the original fume sample, suggesting that material derived from the fume was eliminated in the feces as intact particles. After three days, fecal excretion rates were similar to lung clearance rates for SMAW and GMAW of mild steel.

Chromium and nickel from fumes generated by SMAW of stainless steel were excreted rapidly into the urine during the first few days after instillation, and then at a much slower and gradually diminishing rate. The rapid urinary excretion during the first day was attributed to material that became solubilized in the saline solution used as the vehicle for fume particles during the instillation procedure. After the first day, there was a marked difference in clearance rates of fumes from SMAW and GMAW of stainless steel. SMAW fumes cleared from the lungs with half-times of 73 days for iron, 53 days for chromium, and 49 days for nickel. In contrast, the concentrations of GMAW fumes in the lung remained practically constant throughout the 2-month observation period following exposure (Figure 5). The investigators concluded that a substantial part of the fumes from SMAW of stainless steel

Table 4
Toxicity of Welding Fumes in Rats

	Electrode					
	E70S-3	E308-16	E7018	E6010	E70T-1	E5356
Shield gas	CO2	—	—	—	CO2	Argon
Workpiece	Steel	Steel	Steel	Steel	Steel	Aluminum
IN VIVO:						
Pathology/lung	-	+++	+	+	+	-
Pathology/kidney	-	-	-	-	+	-
Macrophages:						
Decr. phagocytosis	+	+	+	+	+	N.E.
Number of cells in lavage fluid	incr	incr	decr	?	decr	incr
S. typhimurium	-	-	-	-	-	-
Chromosome aberration	-	-	-	-	-	-
IN VITRO:						
Macrophages:						
Viability	high	very low	low	low	low	high
Decr. phagocytosis*	N.E.	13	39	1089	118	359
ATP*	359	4	4	359	359	1089

? = Test not performed
N.E. = No effect
+++ = Severe response
* Lowest fume concentration (ug/ml) that depressed phagocytosis or ATP levels
Data from Coate, Ref. 23

are dissolved in the lung and eventually excreted into urine, whereas fumes from GMAW are less readily dissolved and remain in the lungs for an extended period of time.

Another study by the same research group focused on clearance of manganese from the respiratory tract (Ref. 77). Rats were exposed by inhalation (1 hour/day, 5 days/week for 1 to 4 weeks) to fumes from SMAW of stainless and mild steel and GMAW of stainless steel; the fumes contained 2.2, 2.8, and 7.9% manganese, respectively. Manganese in fumes from SMAW of mild steel cleared rapidly from the lung during the first week after exposure (half-time = 0.5 day) and much more slowly thereafter (half-time = 5 days). Clearance of manganese from fumes generated by SMAW and GMAW of stainless steel was much slower. Manganese from SMAW fumes cleared from the lungs with a half-time of 5 days during the first day and a half-time of 40 days thereafter. The half-time of manganese from GMAW fumes was 107 days throughout the observation period. The clearance of manganese from SMAW fumes paralleled the clearance of iron and the ratio of iron to manganese remained constant throughout the 106-day post-exposure observation period. This suggested that manganese was cleared from the lungs as intact particles and that little manganese clearance resulted from in situ dissolution of particles.

Using transmission electron microscopy and X-ray microanalysis, Anttila demonstrated that the marked differences between pulmonary clearance rates of fumes generated by GMAW and SMAW of stainless steel are related to differences in the solubility of particles in the lung (Refs. 6 and 7). The morphology and chemical composition of particles were examined periodically after exposure of rats to welding fumes by inhalation. Shortly after exposure, all particles were found in alveolar macrophages and type-I epithelial cells; none were found free in the interstitium. Two types of SMAW particles were seen in freshly collected fumes and in the lung. The major population consisted of irregular shaped particles, enriched with sodium and potassium; the minor population contained a higher percentage of iron, nickel, and chromium. The major particle population changed with time, losing practically all of its sodium, potassium, chromium, and also some manganese. By 57 days, so much material was lost that the particles became transparent. In

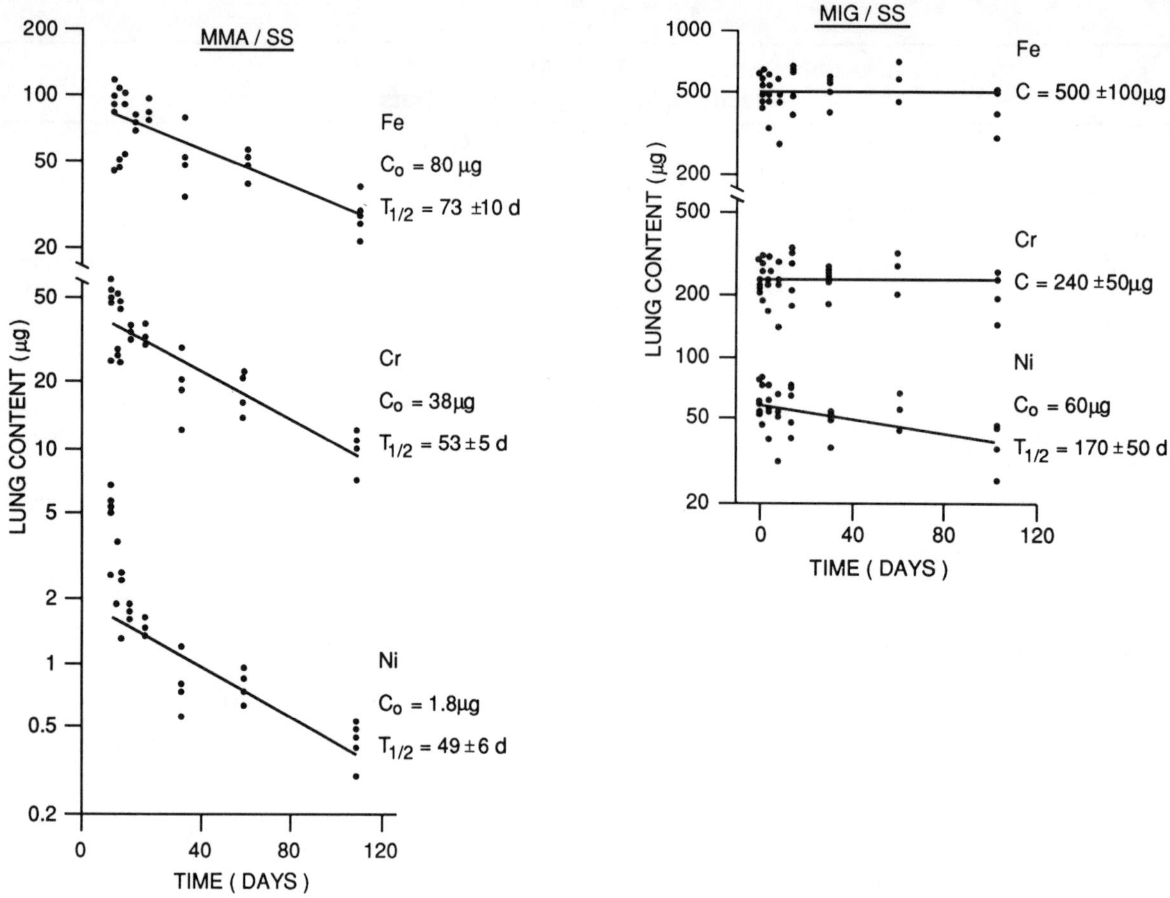

Note: For elements whose levels were decreased in the lung during the 120 day observations period, CO = initial concentration and T1/2 = half time. The average lung content (C) is given for elements that underwent no apparent change in concentration.

Figure 5 — Amounts of Iron, Chromium and Nickel in Rat Lungs After Intratracheal Instillation of Activated Fumes Generated by SMAW and GMAW of Stainless Steel (MMA/SS and MIG/SS, respectively). Kalliomaki et al., Ref. 73

contrast, the minor population was quite stable and lost only sodium and potassium. The particles in GMAW fumes resembled the minor population from SMAW fumes and, like those in SMAW fumes, showed no signs of dissolution during the 10-day study period. The rate of dissolution of the two populations of particles in the lungs reflected the differences in clearance rates described in the above studies. Thus, the solubility of particles may be an important determinant of the pulmonary clearance of particles from welding fumes.

20.3 Toxicity. A long-term inhalation study of the effects of nickel oxide and arsenic trioxide was conducted by Glaser et al. (Ref. 42). Male Wistar rats were exposed continuously for 18 months to aerosols of nickel oxide and arsenic trioxide (particle size $<$ 0.3 MMAD). Two dose levels, 60 ug/m^3 and 200 ug/m^3, were used. Exposure to arsenic trioxide had no apparent effects; there were no significant differences in body weight, hematology, clinical chemistry, gross pathology, or histopathology between treated and control animals. Rats treated with nickel oxide had an increased mortality rate and severe alveolar proteinosis. The red blood cell count, hematocrit, and hemoglobin level were significantly elevated after 9 months of exposure. No treatment-related tumors were seen in either group, but the investigators concluded that survival was too short to assess the carcinogenicity of nickel oxide.

Gorban et al. examined the effects of exposure to fumes generated by CO_2-shielded GMAW of medium and high alloy steel with two types of austenitic electrodes (Ref. 45). The concentrations of chromium, nickel, and manganese were 22%, 7%, and 1.5% for the first electrode and 20%, 9%, and 7% for the second electrode. The LD50s, determined by intraperitoneal injection of fumes, did not differ significantly for the two electrodes (electrode 1: 2167 mg/kg; electrode 2: 2017 mg/kg). Differences noted in the effects of the two welding fumes on levels of alanine and aspartate aminotransferase and nucleic acid levels in the liver and lung were attributed to differences in the content and chemical characteristics of manganese and chromium in the welding fumes.

In other studies by this research group, Gorban et al. (Refs. 43 and 87) determined the acute toxicity of welding fume containing 29.9% manganese, 3.4% nickel, 7.5% trivalent chromium, 0.03% hexavalent chromium, and 24% iron. LD50s, determined by intraperitoneal injection in rats and mice, were 1008 and 1450 mg/kg, respectively. No signs of fibrosis resulted from intratracheal instillation of 50 mg welding fume. Dose-related changes in serum transaminases, hepatic succinate dehydrogenase, and liver and kidney weights were observed in rats following exposure by inhalation for 4 months. From these studies, a maximum permissible exposure of 0.3 mg/m^3 was recommended based on total fume or 0.1 mg/m^3 based on the manganese content of the fume.

20.4 Allergic Sensitivity. Repeated skin contact with chromium salts can produce allergic dermatitis. Caldas and Hicks reported that exposure of the respiratory tract to chromate or extracts of fumes rich in chromium and nickel from SMAW of stainless steel can inhibit this reaction (Refs. 18 and 60). Guinea pigs receiving a series of dermal injections of chromate developed dermal hypersensitivity reactions in response to further dermal challenge with chromate. This response was significantly inhibited in animals given repeated intrapulmonary injections of potassium chromate or potassium dichromate before the dermal sensitization procedure. Pretreatment of the respiratory tract with nickel sulfate did not alter the dermal response to chromate indicating that pulmonary exposure to chromate provokes a specific tolerance to the immunologic effects of chromate.

20.5 Biochemical Studies. Geleskul et al. used the formation of malondialdehyde in the presence of ascorbic acid as a measure of lipid peroxidation produced by welding fumes (Ref. 41). SMAW fume samples were administered to rats by intraperitoneal injection or intratracheal instillation. Fume samples contained 19.8% silicon, 17.3% manganese, and 24.2% iron. Malondialdehyde concentrations were elevated in the lung, kidney, and liver. The effects were greater after intratracheal instillation than after intraperitoneal injection, and the response in the lung was greater than that in the other two organs. The response reached a peak within 24 hours in the lung and within 7 days in the liver and kidney. The investigators concluded that lipid peroxidation can be used to assess the toxicity of welding fumes.

Calmodulin, a calcium-binding protein which regulates diverse cellular processes, is normally activated by calcium, but can also be activated by other heavy metals with a similar ionic radius (e.g., lanthanum, cadmium, mercury, and lead). MacNeil et al. explored the possibility that chromium may exert its toxicity at the subcellular level by interacting with calmodulin (Refs. 29 and 93). Interactions with calmodulin were tested by assaying the effect of metals on beef heart phosphodiesterase, a calmodulin-dependent enzyme. With the exception of chromium, micromolar doses of calcium and the heavy metals tested were required to achieve maximal activation of calmodulin. In contrast, trivalent chromium was only slightly effective in micromolar doses, but nanomolar concentrations of chromium were uniquely effective in activating calmodulin. The investigators suggested that the sensitivity of calmodulin to these low concentrations of chromium could be relevant to the mechanism of chromium toxicity. Because hexavalent chromium is converted to the trivalent form once inside the cell, a similar effect could ultimately be produced by both valence states of chromium.

Red blood cell chromium levels were determined in 22 stainless steel welders who reported no physical symptoms of chromium toxicity and in 11 controls who had no occupational exposure to chromium. The average red blood cell chromium concentration for welders was 28.2 nM compared with 7.5 nM for controls. Because red blood cell concentrations of chromium were of the same order of magnitude as those which activate calmodulin in vitro, the investigators suggested that chromium may exert toxic effects within the cell by interacting with regulatory proteins. However, as the investigators pointed out, chromium in very low levels is an essential element. Thus, it would also seem possible that activation of calmodulin by chromium may be a normal cell function.

20.6 Barium. In 1984, Dare et al. reported that water-soluble barium salts can be extracted from fumes from three types of electrodes containing barium fluxes (Ref. 133). Using the same extraction procedures, Hicks et al. later showed that water-soluble components of fumes from the Soudefonte B1 flux-coated electrode and aqueous solutions of pure barium chloride produce similar physiological effects (Ref. 61). Bronchoconstriction, cardiac arrhythmias, and transient hypertension were observed when solutions of fume extracts or barium chloride were administered to guinea pigs either by intravenous injection or by aerosols introduced into the lungs through tracheal cannulae.

21. In Vitro Studies

21.1 Bacterial Assays. In the past, most studies of the mutagenicity of welding fumes indicated that only stainless steel welding fumes are mutagenic in the Salmonella/Ames test. Unlike previous reports, Biggart et al. reported that the particulate fraction of emissions from SMAW of mild steel are mutagenic when the Salmonella/Ames assay is modified to include preincubation of tester cells in liquid medium with the fume (Ref. 133). No mutagenic activity was observed when the tester bacterial cells were plated with fume particles according to standard protocols in the absence of a preincubation period. The particles contained both direct-acting and promutagenic compounds (promutagens require activation by mammalian metabolic enzymes before mutagenicity is expressed) which induced frameshift mutations.

The direct-acting mutagenic activity was nonspecifically inactivated by microsomal enzymes (i.e., both native and heat-denatured microsomal enzymes inhibited mutagenic activity). Based on these data, and evidence that the mutagenic activity is (1) not altered by incubation with ethylene diamine tetraacetic acid (EDTA), is (2) not extractable with organic solvents, and (3) resides in particles too large to pass through the bacterial cell wall, Biggart et al. concluded that the mutagen(s) associated with welding fume particles are not present in ionic forms and are strongly associated with the particles. They postulated that the fume particle is degraded at the cell surface enabling particle components to be taken up by the bacteria (Ref. 14).

Mutagenic agents were demonstrated in the gas fraction by passing filtered welding aerosol into a chamber containing agar plates seeded with tester cells (gas protocol). This method was far more effective than bubbling the filtered gas through a liquid medium containing bacterial cells (aqueous protocol). Exposure of the tester strains to less than 0.5% of the gas generated from one welding rod induced a 103-fold increase in base-pair substitution mutations with the gas protocol, but only a two-fold increase with the aqueous protocol. Unlike the particulate fraction of welding fumes, exposure to the gas fraction greatly increased base-pair substitutions, but not frameshift mutations. The mutagenicity of the gas fraction was, at most, only partially due to NO_2 since quantities of NO_2 comparable to that present in the welding aerosol produced a substantially lower mutation rate (Ref. 13).

Ong et al. demonstrated that the SOS test, which detects gene damage in Salmonella typhimurium by a colorimetric reaction, can be adapted for use in occupational settings (Refs. 124 and 125). Fumes from SMAW of stainless steel, diesel exhaust, and occupational agents known to be mutagenic in the Salmonella/Ames test were collected by impingement directly into medium containing Salmonella typhimurium tester cells. After several hours of exposure in the collection medium, tester cells were removed from the impinger and subjected to standard SOS test procedures. Diesel exhaust and stainless steel welding fumes induced a high response with the SOS test, and the effects were greater without microsomal enzymes than in their presence. The SOS test was slightly more sensitive to diesel fumes, but much more sensitive to welding fumes than the Salmonella/Ames test. Ong et al. did not attempt to explain this finding. However, their observations may have resulted from differences in the collection methods used for the two tests. Fume samples were collected by impingement into aqueous medium for the SOS test but by filtration for the Ames test. As discussed earlier, Pedersen et al. (Ref. 137) have shown that greater quantities of hexavalent chromium are present in fumes collected by impingement than by filtration. This observation could explain the findings of Ong et al. if hexavalent chromium is responsible for activity in the SOS test.

21.2 Mammalian Cell Studies

21.2.1 Sister Chromatid Exchange (SCE). Two studies examined the induction of SCE in cell cultures by welding and cutting aerosols (Refs. 8 and 182). Water-soluble and insoluble fractions of SMAW fumes generated by five electrodes were tested in mitotic delay and SCE assays with cultured Chinese hamster lung cells by Baker et al. (Ref. 8). In all cases, SCE activity was proportional to the hexavalent chromium content in the fumes; trivalent chromium, fluorides, nickel, manganese, or other fume constituents made, at the most, minor contributions to the induction of SCE. However, the hexavalent chromium content could not account for the mitotic delay caused by fume samples, indicating that other components of the fume fractions produced this effect. Soluble fume fractions were much more active than insoluble fractions in both the mitotic delay and SCE assays, presumably because of their greater bioavailability.

In the second study, Valerio et al. tested fumes released during gas cutting of oil tanks. Oil tank surfaces are frequently covered with thick layers of tar or crude oil. When repairs are made by gas cutting, thick smokes are released with high levels of benzo (a)pyrene and other polycyclic aromatic hydrocarbons. The study showed that smokes emitted by gas cutting during refitting operations in oil tankers produce SCE in murine bone marrow cells (Ref. 182).

21.2.2 Cytotoxicity. Alveolar macrophages play an important role in the defense of the lung against foreign materials. However, their activity may also be an important factor in the development of chronic fibrotic lung disease and emphysema. Because of this, there has been considerable interest in the effects of welding fume particles on macrophage viability and phagocytosis (Refs. 23, 64, and 188). Hooftman et al. showed that the cytotoxicity of welding fume particles generated by different welding processes to bovine alveolar macrophages decreases in the following order: SMAW of stainless steel > GMAW of stainless steel > SMAW of mild steel > GMAW of mild steel. The toxicity of SMAW fume particles was comparable to that of equivalent concentrations of hexavalent chromium in K_2CrO_4, suggesting that hexavalent chromium may be responsible for the cytotoxicity of welding fumes to alveolar macrophages (Ref. 64).

The cytotoxicity of fumes from SMAW and GMAW of mild and stainless steel to rat alveolar macrophages was studied by Pasanen et al. (Ref. 135). Macrophages, obtained by lung lavage, were incubated overnight with intact fume particles or their water-insoluble components. Cell viability, as determined by trypan blue exclusion, and release of the enzymes lactate dehydrogenase (LDH) and lysosomal beta-glucuronidase from macrophages were measured. Fumes from SMAW of mild and stainless steel were equally effective in reducing cell viability and were much more effective than those from GMAW. Fumes generated by GMAW of mild steel were less cytotoxic than those from GMAW of stainless steel. Similar to the findings of Hooftman et al., Pasanen found that the viability of cells treated with fumes from GMAW of mild steel differed little from control cells. Removing water-soluble materials by washing with phosphate-buffered saline substantially reduced the cytotoxicity of fumes from SMAW of stainless steel but had little effect on the other fume particles.

The effect of washed and unwashed fumes on the release of LDH followed the same pattern (Figure 6), whereas the effect on beta glucuronidase was similar for washed and intact particles from all welding processes. Potassium chromate and particles from SMAW of stainless steel had equivalent effects. This suggested that the hexavalent chromium present in fumes from SMAW of stainless steel may be responsible for the toxic effects of these particles.

Two research groups compared the cytotoxic effects of fumes from different welding processes on cultured mammalian cells. Hildebrand and Collyn-d'Hooghe examined the effects of welding fume particles on cultured human embryonic pulmonary epithelial cells (Ref. 63). These investigators found that fumes from SMAW of stainless steel were more cytotoxic than fumes from GMAW of stainless steel. Onoprienko (Ref. 126) compared the cytotoxic effects on mouse fibroblast and human epithelial cell cultures of fumes generated by "powdered" electrodes with those generated by basic and acid electrodes. The relative toxicities of five Russian electrodes was determined on the basis of this experiment.

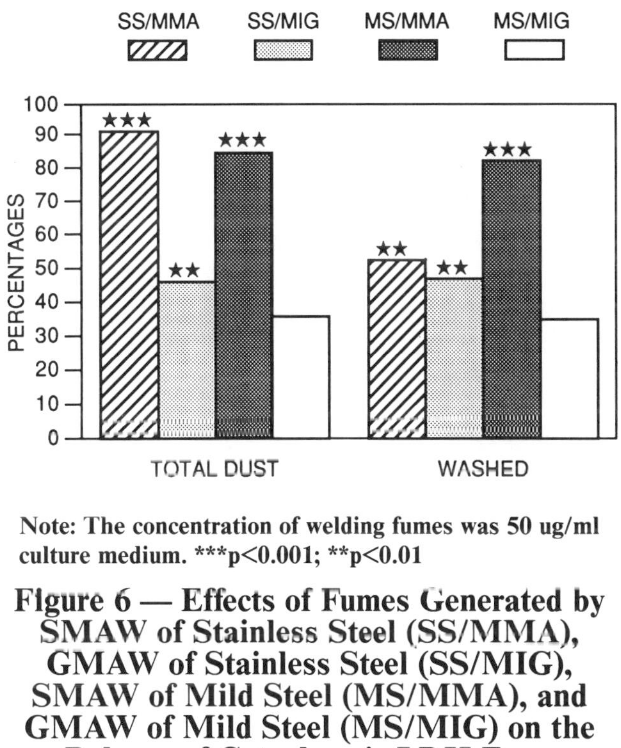

Note: The concentration of welding fumes was 50 ug/ml culture medium. ***p<0.001; **p<0.01

Figure 6 — Effects of Fumes Generated by SMAW of Stainless Steel (SS/MMA), GMAW of Stainless Steel (SS/MIG), SMAW of Mild Steel (MS/MMA), and GMAW of Mild Steel (MS/MIG) on the Release of Cytoplasmic LDH From Cultured Rat Alveolar Macrophages, Before and After Extraction of Water-Soluble Compounds From the Fume Samples Pasanen et al., Ref. 135

21.3 Hyperbaric Pressure. Because of the importance of offshore oil fields in northern Europe, much attention is being paid there to the effects of hyperbaric pressures on divers performing functions such as underwater welding and cutting. Adverse physiological effects can result from exposure to hyperbaric conditions in the absence of chemical exposures. Concern for the effects of fumes created by underwater welding on already stressed physiological systems prompted the study of the combined effects of chromium and high pressures on the cytoskeletal systems of fibroblast cultures (Refs. 68 and 173). The cytoskeletal system, responsible for maintaining cell shape and motility, is composed of chains or filaments of the protein actin. High pressure can disrupt these filaments and alter the shape of the cells causing them to become spherical.

Syversen et al. (Refs. 68 and 173) showed that while 1-hour exposures to either high pressure (100 or 150 bar) or chromate (1 to 5 uM $K_2Cr_2O_7$) caused cells to round up and changed the appearance of the F-actin filaments, the effects were not additive. Thus, the same number of abnormal cells were seen in cultures exposed for one hour to elevated pressures in the absence or presence of chromium. However, the results obtained during 18-hour exposures were quite different. In this case, exposure to 1 uM chromium caused a marked decrease in the number of normal cells exposed to 100 bar pressure, whereas only a small number of cells were affected at normobaric pressures. The synergistic effects seen with 18-hour exposures to 1 uM chromate and 100-bar pressure suggested that the toxicity of chromate may be potentiated by elevated pressures.

In a related study, Jenssen and Syversen studied the effects of chromium and high pressure on cell cycle kinetics in cultured glioma cells (Ref. 68). The cell cycle is divided into four phases in regard to DNA synthesis and cell division: cell division takes place during mitosis (M); the first gap phase (G1) precedes DNA synthesis (S); the second gap phase (G2) precedes cell division. Exposure of cells to 1 uM $K_2Cr_2O_7$ at normobaric pressures (1 bar) caused a prolongation of the S phase, whereas cells exposed to 5 uM $K_2Cr_2O_7$ did not pass through mitosis. High pressure did not alter these effects. However, the combined exposure to chromate and 100-bar pressure resulted in the prolongation of the early part (G1 phase) of the cell cycle. This effect was not seen if cells were exposed to chromate or elevated pressure alone.

References

(1) ACGIH, Threshold limit values and biological exposure indices for 1986-1987. American Conference of Governmental Industrial Hygienists, Cincinnati, OH, (1986).

(2) Anlberg, M. S., Hansson, H. C., Fangmark, I., The growth of a welding fume as a function of relative humidity. In: *Health Hazards and Biological Effects of Welding Fumes and Gases.* Proceedings International Conference, Copenhagen, 18-25 Feb. 1985. Eds: R. M. Stern, A. Berlin, and J. Jarvisalo. Elsevier Science Publishers, Amsterdam, 103-106 (1986).

(3) Aitio A. and Jarvisalo J., Levels of welding fume in tissues and body fluids. In: *Health Hazards and Biological Effects of Welding Fumes and Gases.* Proceedings International Conference, Copenhagen, 18-25 Feb. 1985. Eds: R. M. Stern, A. Berlin, and J. Jarvisalo. Elsevier Science Publishers, Amsterdam, 169-179. (1986).

(4) Angerer, J., Amin, W., Heinrich-Ramm, R., Szadkowski, D., and Lehnert G., Occupational chronic exposure to metals. I. Chromium exposure of stainless steel welders — biological monitoring. *Int. Arch. Occup. Environ. Health, 59,* 503-512 (1987).

(5) Anttila, S. Dissolution of stainless steel welding fumes in the rat lung: An X ray microanalytical study. *Br. J. Ind. Med., 43,* 592-596 (1986).

(6) Anttila, S., Grekula, A., Sivonen S. J., Alapieti, T., Kalliomaki, P. L., and Sutinen, S., Application of transmission electron microscopy and x-ray microanalysis in studying inhaled inorganic particles in-situ in lung tissue. *Third International Conference on Environmental Lung Disease,* Montreal, 15-18 Oct. 1986. *Chest, 91* (2), 313 (1987).

(7) Anttila, S., Sutinen, S., Kalliomaki, P. L., Nickels, J., Grekula, A., and Sivonen, S. J., Structure and composition of inhaled welding fumes in rat lung. In: *Health Hazards and Biological Effects of Welding Fumes and Gases.* Proceedings Int. Conference, Copenhagen, 18-25 Feb. 1985. Eds: R. M. Stern, A. Berlin, and J. Jarvisalo. Elsevier Science Publishers, Amsterdam, 353-356 (1986).

(8) Baker, R. S., Arlauskas, A., Tandon, R. K., Crisp, P. T., and Ellis, J., Toxic and genotoxic action of electric-arc welding fumes on cultured mammalian cells. *J. Appl. Toxicol., 6*(5), 357-362 (1986).

(9) Barlier, A., Guelin, J. C., Jeandel, B., and Salsi, S., Optical radiation in arc welding. Danger and protective measures. *DVS Ber., 105,* ISS Schweissen Schneiden '86, 70-75 (1986).

(10) Baumgardt, B., Jackwerth, E., Otto, H., and Tolg, G., Trace analysis to determine heavy metal load in lung tissue. *Int. Arch. Occup. Environ. Health, 58,* 27-34 (1986).

(11) Beckett, W. S., Davis, J. E., Vroman, N., Nadig, R., and Fortney, S., Heat stress associated with the use of vapor-barrier garments. *J. Occup. Med., 28,* 411-414 (1986).

(12) Benton, R. E., Bonney, T. B., Cole, H.M., and Epstein, S.G., Studies of chemical and physical emissions from welding aluminum alloys. Aluminum Joining Seminar, Metairie, Louisiana, 25-27 Feb. 1986. Publ: The Aluminum Assoc., Washington, D.C., 375-382 (1986).

(13) Biggart, N. W. and Rinehart, R. R., Comparison between aqueous-phase and gas-phase exposure protocols for determining the mutagenic potential of nitrogen dioxide and the gas fraction of welding fumes. *Mutation Res., 188,* 175-184 (1987).

(14) Biggart, N. W., Rinehart, R. R., and Verfaillie J., Evidence for the presence of mutagenic compounds other than chromium in particles from mild steel welding. Mutation Res., 180, 55-65 (1987).

(15) Bjelle, A., Hagberg, M.,and Michaelson, G., Work-related shoulder-neck complaints in industry: A Pilot Study. *Br. J. Rheumatol., 26,* 365-369, (1987).

(16) Bjoerseth, O., Boerresen, E., Malvik, B., and Jakobsen, K., Measurement of welding fume at hyperbaric conditions. In: *Health Hazards and Biological Effects of Welding Fumes and Gases.* Proceedings Int. Conference, Copenhagen, 18-25 Feb.

1985. Eds: R. M. Stern, A. Berlin, and J. Jarvisalo. Elsevier Science Publishers, Amsterdam, 159-162 (1986).

(17) Buskin, S. E. and Paulozzi, L. J., Fatal injuries in the construction industry in Washington State. *Am. J. Ind. Med.*, *11*(4), 453-460, (1987).

(18) Caldas, L. Q. A. and Hicks, R., Intrapulmonary pretreatment by metal-fume components causing inhibition of delayed hypersensitivity. *Arch. Toxicol.*, Suppl. 11, 223-226, (1987).

(19) Callier, A., The truth about contact lenses. *Operations Forum*, 4(2):14; (1987).

(20) Cellini, M., Profazio, V., Fantaguzzi, P., Barbaresi, E., Longanesi, L., and Caramazza, R., Photic maculopathy caused by arc welding. A case report. *Int. Ophthalmol.*, *10*, 157-9 (1987).

(21) Chong, C., Determination of silver, bismuth, cadmium, copper, iron, nickel and zinc in lead- and tin-base solders and white metal bearing alloys by atomic-absorption spectrophotometry. *Talanta*, *33*, 91-94 (1986).

(22) Christensen, J. M. and Kirchhoff, M. Determination of nickel, cadmium and chromium in whole blood and urine by Zeeman absorption spectrophotometry. In: *Health Hazards and Biological Effects of Welding Fumes and Gases.* Proceedings International Conference, Copenhagen, 18-25 Feb. 1985. Eds: R. M. Stern, A. Berlin, and J. Jarvisalo. Elsevier Science Publishers, Amsterdam, 181-184 (1986).

(23) Coate, W. B., Toxicity of Welding fume in rats. Reports: 2236-100/101. Publ: Miami, FL, American Welding Society, (1987).

(24) Cohen, D., Ferromagnetic contamination in the lungs and other organs of the human body. *Science*, *180*, 745-748 (1973).

(25) Corey, P. and Broder, I., Lung cancer in welders. *Third International Conference On Environmental Lung Disease*, Montreal, 15-18 Oct. 1986. *Chest*, *91*(2), 301 (1987).

(26) Cotes, J. E. and El-Gamal, F. M. Cardio-respiratory associations with shipyard welding and burning. In: *Health Hazards and Biological Effects of Welding Fumes and Gases.* Proceedings International Conference, Copenhagen, 18-25 Feb. 1985. Eds: R. M. Stern, A. Berlin, and J. Jarvisalo. Elsevier Science Publishers, Amsterdam, 395-398 (1986).

(27) Cullen, M. R., Cherniack M. G., and Kominsky, J.R. Chronic beryllium disease in the United States. *Seminars in Resp. Med.*, *7*, 203-209 (1986).

(28) Cullen, M. R. Kominsky, J. R. Rossman, M. D., Cherniack, M. G., Rankin, J. A., Balmes, J., Kern, J. A., Daniele, R. P., Palmer, L., Naegel, G. P. McManus, K., Cruz, R. and Cruz, R., Chronic beryllium disease in a precious metal refinery. Clinical epidemiologic and immunologic evidence for continuing risk from exposure to low level beryllium fume. *Am. Rev. Resp. Dis.*, *135*, 201-208 (1987).

(29) Dawson, R., Lakey, T., Morris, B., and Neil, S. M., Activation of calmodulin by chromium its possible relevance to chromium toxicity in man. *5th Joint Meeting of British Endocrine Societies*, Sheffield, England, Apr. 7-10 1986. *J. Endocrinol*, *108* (1986).

(30) De Nevers, N. A., Fatal fire with "nonflammable" methyl chloroform. *Arch. Environ. Health*, *41*, 279-281 (1986).

(31) De Vuyst, P., Dumortier, P., Rickaert, F., Van de Weyer, R., Lenclud, C., and Yernault, J. C. Occupational lung fibrosis in an aluminum polisher. *Eur J. Respir. Dis.*, *68*(2), 131-140 (1986).

(32) Desmeules, M. and Cloutier, L. G., Welder pneumoconiosis; clinical functional and pathological correlations. *Third International Conference on Environmental Lung Disease*, Montreal, 15-18 Oct. 1986. *Chest*, *91*(2), 311, (1987).

(33) Desmeules, M. and Tardif, A., Long-term evolution of pulmonary infiltrates in welders' pneumoconiosis. *Third International Conference On Environmental Lung Disease*, Montreal, 15-18 Oct. 1986. *Chest*, *91*(2), 311, (1987).

(34) Dolan, B. P. and Dolan, D. C., Pulmonary disease survey among union pipefitters/welders in California. In: *Health Hazards and Biological Effects of Welding Fumes and Gases.* Proceedings International Conference, Copenhagen, 18-25 Feb. 1985. Eds: R. M. Stern, A. Berlin, and J. Jarvisalo. Elsevier Science Publishers, Amsterdam, 387-390 (1986).

(35) Donker, B., Gelhaus, J. and Wiesner, P., Hyperbaric welding gas absorber — design and test operation up to 600 m. *Meerestechnik/Marine Technology*, *17*(4), 133-137 (1986).

(36) Drenck, K. and Stern, R. M. Alternating current susceptibility bridge magnetopneumography. In: *Health Hazards and Biological Effects of Welding Fumes and Gases.* Proceedings International Conference, Copenhagen, 18-25 Feb. 1985. Eds: R. M. Stern, A. Berlin, and J. Jarvisalo. Elsevier Science Publisher, Amsterdam, 219-222 (1986).

(37) Ebeid, N. Thiocyanates in Sputum: Relation to welding fumes, tobacco smoking and carcinogenicity. *J. Soc. Occup. Med.*, *37*, 59-61 (1987).

(38) Emmerling, G., Zschiesche, W., Schaller, K. H., Weltle, D., Valentin, H. and Zober, A., Occupational-health study of bronchial hypersensitivity during electric arc welding of surface-coated steels. *DVS Ber.*, *97*, (1987).

(39) Engstrom, B. and Virtamo, M., On site ozone measurements in welding shops. In: *Health Hazards and Biological Effects of Welding Fumes and Gases.* Proceedings International Conference, Copenhagen, 18-25 Feb. 1985. Eds: R. M. Stern, A. Berlin, and J. Jarvisalo. Elsevier Science Publishers, Amsterdam, 163-166 (1986).

(40) Farrell, F. J., Angioedema and urticaria as acute and late phase reactions to zinc fume exposure, with associated metal fume fever-like symptoms. *Am. J. Ind. Med., 12,* 331-337 (1987).

(41) Geleskul, Y. F., Gorban, L. N., Baida, N. A., Voitkevich, V. G. and Kornuta, N. A., Peroxidation of lipids in animal tissues as an index of biological agressiveness of welding aerosols. *Vrach. Delo, 6,* 108-110 (1987).

(42) Glaser, U., Hochrainer, D., Oldiges, H., and Takenaka, S., Long-term inhalation studies with NiO and As_2O_3 aerosols in Wistar rats. In: *Health Hazards and Biological Effects of Welding Fumes and Gases.* Proceedings International Conference, Copenhagen, 18-25 Feb. 1985. Eds: R.M. Stern, A. Berlin, and J. Jarvisalo. Elsevier Science Publishers, Amsterdam, 325-328 (1986).

(43) Gorban, L. N., Prilipkov, A., Cerednicenkov, M., and Kireev, V. I., Bases for a MAC for welding fume containing manganese, chromium and nickel in workplace air. *Gig Tr. Prof. Zabol., 5,* 26-29 (1986).

(44) Gorban, L. N. and Borodiuk, T.M., Hygienic characteristics of exposure of the body to magnesium and aluminum magnide dust and to the aerosol formed during the welding of an aluminum-magnesium alloy. *Vrach Delo,* ISS 2, 105-108 (1986).

(45) Gorban, L. N., Prilipko, V. A., Voitkevich, V. G., and Fedorina, E. E., Particle structure of a welding aerosol and its biological activity. *Gig. Sanit., 7* 23-26 (1986).

(46) Gray, C. N., Hexavalent chromium in welding fume and the role of ozone [letter]. *Ann. Occup. Hyg., 31,* 269-70 (1987).

(47) Grekula, A., Peura, R., and Sivonen, S. J., Quantitative energy dispersive x-ray microanalysis of welding fumes. In: *Health Hazards and Biological Effects of Welding Fumes and Gases.* Proceedings International Conference, Copenhagen, 18-25 Feb. 1985. Eds: R. M. Stern, A. Berlin, and J. Jarvisalo. Elsevier Science Publishers, Amsterdam, 85-88 (1986).

(48) Grekula, A., Ristolainen, E., Tanninen, V. P., Hyvarinen, H. K., and Kalliomaki, P. L., Surface and bulk chemical analysis on metal aerosols generated by manual metal arc welding of stainless steel. *J. Aerosol Science, 17,* 1-9 (1986).

(49) Grekula, A., Sivonen, S. J., Hyvarinen, H. K., Tanninen, V. P., and Kalliomaki, P. L., Analytical electron microscopy of welding fumes. University of Oulu, Reports of the Institute of Electron Optics, Report 1, Oulu, Finland, (1986).

(50) Gustavsson, C. and Welinder, H., Stainless steel manual metal-arc welding and chromium in blood. In: *Health Hazards and Biological Effects of Welding Fumes and Gases.* Proceedings International Conference, Copenhagen, 18-25 Feb. 1985. Eds: R. M. Stern, A. Berlin, and J. Jarvisalo. Elsevier Science Publishers, Amsterdam, 185-188 (1986).

(51) Hagberg, M., Lindqvist, B., and Wall, S., Exposure to welding fumes and chronic renal diseases, a negative case-referent study. *Int. Arch. Occup. Environ., Health, 58*(3), 191-195 (1986).

(52) Hansen, H. S., Sogaard, H., and Jurgensen K. S.-L., The significance of occupational exposure for cancer in the nasal cavity and paranasal sinuses. A study with reference to cases in Denmark, Sweden and Finland. *Ugeskr. Laeg., 148*(34), 2185-2187 (1986).

(53) Heikkila, P. R., Hameila, M., Pyy, L., and Raunu, P., Exposure to Creosote in the impregnation and handling of impregnated wood. *Scand. J. Work Environ. Health, 13,* 431-437 (1987).

(54) Heinrich-Ramm, R. and Angerer, J., Determination of nickel and chromium in body fluids. *Trace elements in human health and disease,* Second Nordic Symposium, Odense, Denmark, 1987. 70-73 (1987).

(55) Heinsalmi, P., Method to measure working posture loads at working sites (OWAS). *The Ergonomics of Working Postures. Models, Methods and Cases,* N. Corlett, J. Wilson, and I. Manenica, Editors, London, Taylor and Francis, 100-104 (1986).

(56) Helander, M. G., Karwan, M. H., and Etherton, J. E., A model of human reaction time to dangerous robot arm movements. *Proceedings of the Human Factors Society, 31st Annual Meeting,* 191-195 (1987).

(57) Helander, M. G. Ergonomics, workplace design. *International Encyclopedia of Robotics Applications and Automation,* John Wiley and Sons, Inc., 477-488 (1988).

(58) Henderson, I. D., Fume generation and chemical analysis of fume for a selected range of flux-cored structural steel wires. AWRA documents P9-44-85 (revised 15/9/86) *Aust. Weld. Res., 15,* 4-11 (1986).

(59) Hewitt, P. J. and Madden, M. G., Welding process parameters and hexavalent chromium in MIG fume. *Ann. Occup. Hyg., 30,* 427-434 (1986).

(60) Hicks, R. and Caldas, L. Q. A., Immunologi-

cal unresponsiveness of chromium or nickel in the guinea-pig induced by stainless steel welding fume components. *Arch. Toxicol.*, Suppl. 9. 421-422 (1986).

(61) Hicks, R., Caldas, L. Q. A., Dare, P. R., and Hewitt, P. J., Cardiotoxic and bronchoconstrictor effects of industrial metal fumes containing barium. *Arch. Toxicol.*, Suppl. 9, 416-20 (1986).

(62) Hicks, R., Oshodi, R. O., and Pedrick, M. J., Cytotoxic, irritant and fibrogenic effects of metal-fume particulate materials investigated by intramuscular injection in the rat and guinea pig. *Arch. Toxicol.*, Suppl. 11, 223-226 (1987).

(63) Hildebrand, H. F., Collyn-d'Hooghe, M., and Stern, R. M. Cytotoxic effects of welding fumes on human embryonic epithelial pulmonary cells in culture. In: *Health Hazards and Biological Effects of Welding Fumes and Gases.* Proceedings International Conference, Copenhagen, 18-25 Feb. 1985. Eds: R. M. Stern, A. Berlin, and J. Jarvisalo. Elsevier Science Publishers, Amsterdam, 319-324 (1986).

(64) Hooftman, R. N., Roza, P., and Arkesteijn, C. W. M., In vitro cytotoxicity of welding fume particles to bovine alveolar macrophages In: *Health Hazards and Biological Effects of Welding Fumes and Gases.* Proceedings International Conference, Copenhagen, 18-25 Feb. 1985. 1985. Eds: R. M. Stern, A. Berlin, and J. Jarvisalo. Elsevier Science Publishers, Amsterdam, 311-314 (1986).

(65) Ingle, J., Device for automatic recording of arcing periods during electric welding. *Ann. Occup. Hyg.*, 30, 123-124 (1986).

(66) Jakubcik, L., Use of RVD-F respirators in welding of high-alloy manganese steels. *Zvaranie, 36*(12), 374-376 (1987).

(67) Jiao, C., Mao, X., Zhang, Z., and Bai, J., Development of model ALJB 1.5-1 welding fume purifier. *Gongye Weisheng Yu Zhiyebing, 12*(3), 179-82 (1986).

(68) Jenssen, J. and Syversen, T., High hydrostatic pressure and chromate-induced effects on cell cycle kinetics. *Undersea Biomed Res, 14*(1), 21-9 (1987).

(69) Jordan, D. R., The potential damaging effects of light on the eye (Part I). *Canadian J. Opthalmology, 21*, 216-224 (1986).

(70) Kalliomaki, P. L., Aitio, A., Hyvarinen, H. K. Kalliomaki, K., and Lakomaa, E. L., Lung clearance, transportation, and excretion of metals in rats after intratracheal installation of activated welding fumes. *In: Health Hazards and Biological Effects of Welding Fumes and Gases.* Proceedings International Conference, Copenhagen, 18-25 Feb. 1985. Eds: R. M. Stern, A. Berlin, and J. Jarvisalo. Elsevier Science Publishers, Amsterdam, 345-348 (1986).

(71) Kalliomaki, P. L., Aitio, A., Lakomaa, E. L., and Kalliomaki, K., Kinetics of the metal components of intratracheally instilled mild and stainless steel welding fumes in rats. *J. Aerosol Sci., 18*(16), 737-740 (1987).

(72) Kalliomaki, P. L., Grekula, A. Hagberg, J., and Sivonen, S., Analytical electron microscopy of welding fumes. *J. Aerosol Sci.*, 18(6), 781-784, (1987).

(73) Kalliomaki, P. L., Hyvarinen, H. K., Aitio, A., Lakoma, E. L., and Kalliomaki, K., Kinetics of the metal components of intratracheally instilled stainless steel welding fume suspensions in rats. *Br J. Ind Med, 43*(2), 112-119 (1986).

(74) Kalliomaki, P. L., Kalliomaki, K., and Korhonen, O., Lung retention of welding fumes and ventilation lung functions — a follow-up study of shipyard welders. In: *Health Hazards and Biological Effects of welding Fumes and Gases.* Proceedings International Conference, Copenhagen, 18-25 Feb. 1985. Eds: R. M. Stern, A. Berlin, and J. Jarvisalo. Elsevier Science Publishers, Amsterdam, 409-412 (1986).

(75) Kalliomaki, K., Kalliomaki, P. L., and Moilanen, M., A mobile magnetopneumorgraph with dust quality sensing. In: *Health Hazards and Biological Effects of Welding Fumes and Gases.* Proceedings International Conference, Copenhagen, 18-25 Feb. 1985. Eds: R. M. Stern, A. Berlin, and J. Jarvisalo. Elsevier Science Publishers, Amsterdam, 215-218 (1986).

(76) Kalliomaki, K., Kalliomaki, P. L., and Moilanen, M., Stability of remanent magnetism of different welding fumes. In: *Health Hazards and Biological Effects of Welding Fumes and Gases.* Proceedings International Conference, Copenhagen, 18-25 Feb. 1985. Eds: R. M. Stern, A. Berlin, and J. Jarvisalo. Elsevier Science Publishers, Amsterdam, 223-226 (1986).

(77) Kalliomaki, P. L., Lakomaa, E. L., Aitio, A., and Kalliomaki, K., Manganese distribution in rats after three types of welding fume exposures. *In: Health Hazards and Biological Effects of Welding Fumes and Gases.* Proceedings International Conference, Copenhagen, 18-25 Feb. 1985. Eds: R. M. Stern, A. Berlin, and J. Jarvisalo. Elsevier Science Publishers, Amsterdam, 349-352 (1986).

(78) Keyserling, W.M., A Computer-aided system to evaluate postural stress in the workplace. *Am. Ind. Hyg. Assoc. J., 47*, 641-649 (1986).

(79) Kiilunen, M., Jarvisalo, J., Makitie, O., and

Aitio, A., Analysis, storage stability and reference values for urinary chromium and nickel. *Int. Arch. Occup. Environ. Health*, *59*, 43-50 (1987).

(80) Kilburn, K. H., Medical screening for lung cancer: perspective and strategy. *J. Occup. Med.*, *28*, 714-718 (1986).

(81) Kilburn, K. H., Respiratory functional impairment in metal welders: An interpretation of current data. In: *Health Hazards and Biological Effects of Welding Fumes and Gases.* Proceedings International Conference, Copenhagen, 18-25 Feb. 1985. Eds: R. M. Stern, A. Berlin, and J. Jarvisalo. Elsevier Science Publishers, Amsterdam, 363-378 (1986).

(82) Kilburn, K. H., Warshaw R., Boylen C. T., Airway obstruction, volume loss, and respiratory symptoms in shipyard welders in Los Angeles: Baseline values and cross shift changes. In: *Health Hazards and Biological Effects of Welding Fumes and Gases.* Proceedings International Conference, Copenhagen, 18-25 Feb. 1985. Eds: R. M. Stern, A. Berlin, and Jarvisalo. Elsevier Science Publishers, Amsterdam, 383-386 (1986).

(83) Kjuus, J., Skjaerven, R, Langard, S., Lien, J. T., and Aamodt, T., A case-referent study of lung cancer, occupational exposures and smoking. *Scand. J. Work Environ. Health*, *12*, 193-202 (1986).

(84) Krasnyuk, E. P., Lubyanova I. P., and Timofeeva, N. T., Changes in the respiratory and cardiovascular systems of alloyed steel welders. In: *Health Hazards and Biological Effects of Welding Fumes and Gases.* Proceedings International Conference, Copenhagen, 18-25 Feb. 1985. Eds: R. M. Stern, A. Berlin, and J. Jarvisalo. Elsevier Science Publishers, Amsterdam, 419-421 (1986).

(85) Krause, H. J. and Press, R.H., Measurement of the welding fume quantities and analysis of the fume composition during resistance welding. *In: Health Hazards and Biological Effects of Welding Fumes and Gases.* Proceedings International Conference, Copenhagen, 18 25 Feb. 1985. Eds: R. M. Stern, A. Berlin, and J. Jarvisalo. Elsevier Science Publishers, Amsterdam, 65-75 (1986).

(86) Krizmanic-Mrkoci, D., Lovric, M., Brkic, K., Ilic, Z., and Mitrecic, A., Dystrophia retinae pigmentosa sine pigmento as a cause of early working disability. *Arh. Hig. Rada Toksikol (Yugoslavia)*, *38*(2), 159-162 (1987).

(87) Kundiev, Y. and Gorban, L. N., Some approaches to the establishment of permissible levels of welding aerosols in the air. In: *Health Hazards and Biological Effects of Welding Fumes and Gases.* Proceedings International Conference, Copenhagen, 18-25 Feb. 1985. Eds: R. M. Stern, A. Berlin, and J. Jarvisalo. Elsevier Science Publishers, Amsterdam, 583-586 (1986).

(88) Landsberger, S., Non-destructive determination of zinc in biological certified reference materials. *J. Radioanal. Nucl. Chem., Letters*, *127*(5), 319-323 (1988).

(89) Landsberger, S. and Simsons, A., Chromium, nickel, and arsenic determinations in human samples by thermal and epithermal neutron activation analyses. *Biol. Trace Elem. Res.*, *13*, 357-362 (1987).

(90) Lerchen, M. L., Wiggins, C. L., and Samet, J. M., Lung cancer and occupation in New Mexico. *J. Natl. Cancer Inst.*, *79*, 639-645 (1987).

(91) Lippmann, M., Magnetopneumography as a tool for measuring lung burden of industrial aerosols. In: *Health Hazards and Biological Effects of Welding Fumes and Gases.* Proceedings International Conference, Copenhagen, 18-25 Feb. 1985. Eds: R. M. Stern, A. Berlin, and J. Jarvisalo. Elsevier Science Publishers, Amsterdam, 199-213 (1986).

(92) Lyngenbo, O., Groth, M. V., Groth, S., Olsen, O., and Dirksen, H., Respiratory symptoms and pulmonary function among Danish welders. In: *Health Hazards and Biological Effects of Welding Fumes and Gases.* Proceedings International Conference, Copenhagen, 18-25 Feb. 1985. Eds: R. M. Stern, A. Berlin, and J. Jarvisalo. Elsevier Science Publishers, Amsterdam, 399-402 (1986).

(93) MacNiel, S., Dawson, R., Lakey, T., and Morris, B., Activation of calmodulin by the essential trace element chromium. *Cell Calcium*, *8*(3) 207-216 (1987).

(94) Maiti, B. and Desai, S. R., High-performance liquid chromatographic separation of beryllium, cobalt, nickel and chromium as the beta-isopropyltropolone complexes and its application to the determination of chromium in air. *Analyst*, *111*(7), 809-811 (1986).

(95) Mastromatteo, E., Nickel. *Am. Ind. Hyg. Assoc. J.*, *47*(10, 589-601 (1986).

(96) McIlwain, J. F. and Neumeier, L. A., Fumes from shielded metal arc welding electrodes. *Govt Reports Announcements & Index (GRA&I)*, Issue 07 (1987).

(97) McLaughlin, J. K., Malker, H. S. R., Blot, W. J., Malker, B. K., Stone, B. J., Weiner, J. A., Ericsson, J. L. E., and Fraumeni, J. F., Occupational risks for intracranial gliomas in Sweden. *J. Natl. Cancer Inst.*, *78*(2), 253-257 (1987).

(98) McLaughlin, J. K., Thomas, T. L., Stone, B. J., Blot, W. J., Malker, H. S. R., Stone, B. J., Weiner,

J. A., Ericsson, J. L. E., and Malker, B.K., Occupational risks for intracranial gliomas in Sweden. *J. Occup. Med.*, *28*(1), 66-68 (1987).

(99) McMillan, G. H. G., Foiling flashes. *Occup. Health (Lond.)*, *38*(1), 20-22 (1986).

(100) McMillan, G. H. G., Metal fume fever. *Occup. Health (Lond.)*, *38*(5), 148-149 (1986).

(101) Moller, D. R., Brooks, S. M., Bernstein, D. I., Cassedy, K., Enrione, M., and Bernstein, I. L., Delayed anaphylactoid reaction in a worker exposed to chromium. *J. Allergy Clin. Immunol.*, *77* (3), 451-456 (1986).

(102) Moravik, J., Malik, E., Morihladkova, V., and Papayova, A., Influence of welding fumes on the respiratory tract of welders. *Cesk. Hyg.*, *31*(2), 77-83 (1986).

(103) Moreton, J., The case for standardization of tests for welding fume emission rate and chemical composition determination. *Ann. Occup. Hyg.*, *30*, 435-444 (1986).

(104) Mylius, E. A. and Gullvag, B., Alveolar Macrophage Count as an Indicator of Lung Reaction to Industrial Air Pollution. *Acta Cytologica*, *30*, 157-162 (1986).

(105) Najarian, T. and Colton, T., Mortality from leukemia and cancer in shipyard nuclear workers. *Lancet*, *1*, 1018-1020 (1978).

(106) Newcova, A., Use of x-ray fluorescence analysis in the study of welding aerosols. *Zvaranie*, *35*(7), 215-219 (1986).

(107) Nilsson, B., Taste acuity of the human palate. *Acta Odontol. Scand.*, *37*, 217-234 (1979).

(108) Nordman, H., Keskinen, H., Savolaninen, A., and Rahkonen, E., Asthmatic reactions in stainless steel welders. *Third International Conference on Environmental Lung Disease*, Montreal, 15-18 Oct. 1986. *Chest*, 91(2), 311 (1987).

(109) Norell, S., Ahlbom, A., Olin, R., Erwald, R., Jacobson, G., Lindberg-Navie S., and Wiechel, K. L., Occupational factors and pancreatic cancer. *Br. J. Ind. Med.*, *43*, 775-778 (1986).

(110) Norseth, T., The carcinogenicity of chromium and its salts. *Br. J. Ind. Med.*, *43*(10), 649-651 (1986).

(111) Norseth, T., Metal carcinogenicity. In: *Living in a Chemical World. Ann. N.Y. Acad. Sci.*, *534*, 377-385 (1988).

(112) Ohlsson, A. C., Jarvholm, B., Lofqvist, A., Naslund, P. E., and Stenborg, R., Vocal behaviour in welders-a preliminary study. *Folia Phoniatr (Basel)*, *39*(2), 98-103 (1987).

(113) Okuno, T., Measurement of blue-light effective radiance of welding arcs. *Ind Health*, *24*(4), 213-226 (1986).

(114) Okuno, T., Measurement of ultraviolet radiation from welding arcs. *Ind. Health*, *25*, 147-156 (1987).

(115) Olah, L., Hygienic aspects of chromium and nickel occurrence in the welding working environment. *Cesk. Hyg.*, *32*(4), 225-231 (1987).

(116) Olah, L., Hygienic characteristics of hardfacing consumables. *Zvaracske Spravy*, *37*(3), 64-72 (1987).

(117) Olah, L., The use of nuclear analytical methods for the determination of heavy metals in the working environment of welding operators. *Prac. Lek*, *38*(1), 9-14 (1986).

(118) Olah, L.,and Guldan, P., Hygienic study of welding with VUZ TNE 800 and VUZ A6 hardfacing electrodes. *Zvaranie*, *35*(5), 156-159 (1986).

(119) Olah L. and Mihalkovic, K., Hygienic evaluation of flux-cored welding rods for surfacing. *Zvaranie*, *36*(2), 57-60 (1987).

(120) Olah, L. and Pivoluska, J., Method for determination of chromium (VI) in the working environment in industrial processing of high-alloy steels. *Hutn. Listy*, *41*(4), 287-289 (1986).

(121) Olander, L., Welding fume plumes in industrial workrooms. *Chem. Eng. Monogr.*, *24*, ISS Ventilation '85, 153-158 (1986).

(122) Oleru, G. and Ademiluyi, S. A., Some acute and long-term effects of exposure in welding and thermal-cutting operations in Nigeria. *Int. Arch. Occup. Environ. Health*, *59*, 605-612 (1987).

(123) Olin, R., Anlbom, A., Lindberg-Navie, S., Norell, S., and Spannare, B., Occupational factors associated with astrocytoma: a case control study. *Am. J. Ind. Med.*, *11*, 615-625 (1987).

(124) Ong, T., Stewart, J., Wen, Y. F., and Whong, W. Z., Application of sos umu-test for the detection of genotoxic volatile chemicals and air pollutants. *Environ. Mutagenesis*, 9(2), 171-176 (1987).

(125) Ong, T., Stewart, J. D., and Whong, W. Z., Induction of sos responses in Salmonella Typhimurium by volatile chemicals and environmental pollutants. *Environ. Mutagen*, *8*(6), 62 (1986).

(126) Onoprienko, E. N. and Suprun, S.A., Hygienic characteristics of flux-cored welding rods. *Metall. Tekhnol. Probl. Svarki Poroshk. Provolokoi*, Dokl. Mezhdunar. Shk. Stran-Chlenov SEV, 104-108 (1986).

(127) Ortendahl, T., Oral changes in divers working with electrical welding/cutting underwater. *Swedish Dental J.*, Suppl. 43, 51 (1987).

(128) Ortendal, T. W. and Hogstedt, P., Magnetic fields effects on dental amalgam in divers welding and cutting electrically underwater. *Undersea Biomed. Res.*, in press (1988).

(129) Ortendal, T. W., Hogstedt, P., Odelius, H.,

and Noren, J. G., Effects of magnetic fields from underwater electrical cutting on in vitro corrosion of dental amalgam. *Undersea Biomed. Res.*, in press (1988).

(130) Ortendahl, T. W. and Holland, R.I., Influence of submerged electrical welding/cutting on dental amalgam in Swedish divers. *Scand. J. Dent. Res.*, *94*(6), 545-52 (1986).

(131) Ortendahl, T. W. and Holland, R.I., Electrical activity in dental amalgam of submerged divers during welding. *Acta Odontol. Scand* ., *45*, 313-20 (1987).

(132) Palmer, W. G., *Effects of Welding on Health IV.* Miami, FL, American Welding Society, 1983.

(133) Palmer, W. G., *Effects of Welding on Health VI.* Miami, FL American Welding Society, 1989.

(134) Parsons, H. M., Robotics and the health of workers. In: *Occupational Health and Safety in Automation and Robotics*, Proceedings 5th UOEH International Symposium, Kitakyushu, Japan, 20-21 Sep. 1985. Ed.: K. Noro, New York, Taylor and Francis, 196-211 (1987).

(135) Pasanen, J. T., Gustafsson, T. E., Kalliomaki, P. L., Tossavainen, A., and Jarvisalo, J. O., Cytotoxic effects of four types of welding fumes on macrophages in vitro: A comparative study. *J. Toxicol. Environ. Health*, *18*, 143-152 (1986).

(136) Pearce, N. E. and Howard, J. K., Occupation, social class and male cancer mortality in New Zealand, 1974-78. *Int. J. Epid.*, *15*, 456-462 (1986).

(137) Pedersen, B., Thomsen, E., and Stern, R. M., Some problems in sampling, analysis and evaluation of welding fumes containing Cr(VI). *Ann. Occup. Hyg.*, *31*, 325-338 (1987).

(138) Pekkarinen, J., Industrial impulse noise, crest factor and the effect of earmuffs. *Am. Ind. Hyg. Assoc. J.*, *48*, 861-866 (1987).

(139) Peto, J., Cancer morbidity and mortality studies of welders. In: *Health Hazards and Biological Effects of Welding Fumes and Gases.* Proceedings International Conference, Copenhagen, 18-25 Feb. 1985. Eds: R. M. Stern, A. Berlin, and J. Jarvisalo. Elsevier Science Publishers, Amsterdam, 423-434 (1986).

(140) Pierre, F., Boudier, C., Pelletier, A., Salsi, S., Vigneron, J. C., and Bieth, J.G., Effect of welding fumes on lung antiproteases. *Inst. Natl. Sante Rech. Med.*, [Colloq.], *155*, ISS Silicose Pneumoconioses Poussieres Mixtes, 129-33 (1987).

(141) Puntoni, R., Vercelli, M., Digiorgio, F., Valerio, F., Bonassi, S., Ceppi, M., and Santi, L., Mortality study among autogenous and electrical welders [gas and arc welders] in the port of Genoa (Italy). In: *Health Hazards and Biological Effects of Welding Fumes and Gases.* Proceedings International Conference, Copenhagen, 18-25 Feb. 1985. Eds: R. M. Stern, A. Berlin, and J. Jarvisalo. Elsevier Science Publishers, Amsterdam, 469-472 (1986).

(142) Randolph, S. A. and Zavon, M. R., Guidelines for contact lens use in industry. *J. Occup. Med.*, *29*(3), 237-242 (1987).

(143) Rasmussen, K. H., and Jepsen, J. R., The organic psychosyndrome in manual metal arc welders: A possible consequence of manganese toxicity. *UGESKR LAEG*, *149*(51), 3497-3498 (1987).

(144) Reith, A., Voss, R., and Reichborn-Kjennerud, S., A simple atraumatic sampling method for screening of precancerous and toxicological lesions by cytology in the upper respiratory tract. In: *Health Hazards and Biological Effects of Welding Fumes and Gases.* Proceedings International Conference, Copenhagen, 18-25 Feb. 1985. Eds: R. M. Stern, A. Berlin, and J. Jarvisalo. Elsevier Science Publishers, Amsterdam, 591-594 (1986).

(145) Roeder, M., Development of a mobile apparatus for detection and removal of welding fumes in a plant with recirculating air. *Tech. Umweltschutz*, *33*, ISS Umweltgestaltung produktionsbereichen, 116-19 (1986).

(146) Rosen, G., and Lundstrom, S., Concurrent video filming and measuring for visualization of exposure. *Am. Ind. Hyg. Assoc. J.*, *48*, 688-692 (1987).

(147) Rudell, B., Kolmodin-Hedman, B., and Wenngren B. I., Controlled welding in patients with neurological symptoms. In: *Health Hazards and Biological Effects of Welding Fumes and Gases.* Proceedings International Conference, Copenhagen, 18-25 Feb. 1985. Eds: R. M. Stern, A. Berlin, and J. Jarvisalo. Elsevier Science Publishers, Amsterdam, 413-414 (1986).

(148) Sawatari, K. and Serita, F., Determination of chromium speciation in fumes prepared by a plasma metal sprayer as a model of actual welding fumes. *Ind. Health*, *24*, 51-61 (1986).

(149) Schneider, W. D., Liebich, R., Maintz, G., Schmidt, G., and Wiesener, W., Exposure and state of health in electric arc welders using different technologies. In: *Health Hazards and Biological Effects of Welding Fumes and Gases.* Proceedings International Conference, Copenhagen, 18-15 Feb. 1985. Ed: R. M. Stern, A. Berlin, and J. Jarvisalo. Elsevier Science Publisher, Amsterdam, Netherlands, 391-394 (1986).

(150) Schneider, W. D., Maintz, G., Reimer, W., Schmidt, G., and Tittelbach, U., Are there functional hints of fibrosis in arc welders' siderosis. *Clin. Respir. Physiol.*, *23*(12), 3498 (1987).

(151) Schneider, W. D., Maintz, G., Reimer, W., Schmidt, G., and Tittelbach, U., Functional signifi-

cance of lung siderosis in electrode welders. *Z Gesamte Inn. Med.*, 42(5), 126-130 (1987).

(152) Schoenberg, J. B., Stemhagen, A., Mason, T. J., Patterson, J., Bill, J., and Altman, R., Occupation and lung cancer risk among New Jersey white males. *J. Natl. Cancer Inst.*, *79*(1), 13-21 (1987).

(153) Sharir, M. and Ben-David, A., Severe Keratoconjunctivitis in plasma flame-spraying workers [letter]. *J. Occup. Med.*, *29*, 639 (1987).

(154) Shehade, S. A., Roberts, P. J., Diffey, B. L., and Foulds, I. S., Photodermatitis Due to Spot Welding. *Br. J. Dermatology*, *117*, 117-119 (1987).

(155) Shusterman, D. and Neal, E., Skeletal muscle and myocardial injury associated with metal fume fever. *J. Fam. Pract.*, *23*(2), 159-160 (1986).

(156) Silberschmid, M., Acute pneumonitis progressing to pulmonary fibrosis in a welder. In: *Health Hazards and Biological Effects of Welding Fumes and Gases.* Proceedings International Conference, Copenhagen, 18-25 Feb. 1985. Eds: R. M. Stern, A. Berlin, and J. Jarvisalo. Elsevier Science Publishers, Amsterdam, 415-418 (1986).

(157) Sipek, L. and Smars, E., Ozone and nitrogen oxides in gas shielded arc welding. In: *Health Hazards and Biological Effects of Welding Fumes and Gases.* Proceedings International Conference, Copenhagen, 18-25 Feb. 1985. Eds: R. M. Stern, A. Berlin, and J. Jarvisalo. Elsevier Science Publishers, Amsterdam, 107-110 (1986).

(158) Sjogren, B. and Carstensen, J., Cancer morbidity among Swedish welders and gas cutters. In: *Health Hazards and Biological Effects of Welding Fumes and Gases.* Proceedings International Conference, Copenhagen, 18-25 Feb. 1985. Eds: R. M. Stern, A. Berlin, and J. Jarvisalo. Elsevier Science Publishers, Amsterdam, 461-463 (1986).

(159) Sjogren, B., Elinder, C. G., Lidums, V. and Changs, G., Uptake and urinary excretion of aluminum by welders. *Arbete Och Halsa*, *16* (1986).

(160) Sjogren, B., Elinder, C. G., Lidums, V., and Changs, G., Uptake and urinary excretion of aluminum among welders. *Int. Arch. Occup. Environ. Health*, *60*, 77-79 (1988).

(161) Sjogren, B., Gustavsson, A., and Hedstrom, L., Mortality in two cohorts of welders exposed to high and low-levels of hexavalent chromium. *Scand. J. Work Environ. Health*, *13*, 247-251 (1987).

(162) Sjogren, B., Weiner, J., Horte, L. G., and Carstensen J. Mortality amongst Swedish welders and gas cutters. In: *Health Hazards and Biological Effects of Welding Fumes and Gases.* Proceedings International Conference, Copenhagen, 18-25 Feb. 1985. Eds: R. M. Stern, A. Berlin, and J. Jarvisalo. Elsevier Science Publishers, Amsterdam, 457-459 (1986).

(163) Sliney, D. H., Benton, R. E., Cole, H. M., Epstein, S. G., and Morin, C.J., Transmission of potentially hazardous actinic ultraviolet radiation through fabrics. *Appl. Ind. Hyg.*, *2*, 36-44 (1987).

(164) Smith, N. J., Topping, M. D., Stewart, J. D., and Fletcher, J. G., Occupational cadmium exposure in jig solderers. *Br. J. Ind. Med.*, *43*(10), 663-666 (1986).

(165) Spacilova, M. and Hykes, P., General health state of electrowelders and persons at risk of free silicon oxide with similar X-ray picture. *Prac. Lek.*, 38(9), 389-392 (1986).

(166) Spacilova, M. and Hykes, P., The evolution of pulmonary changes of electrical welders in relation to work risk and smoking habits. *Prac. Lek.*, *38* (6), 234-239 (1986).

(167) Stage, J. and Vinding, T., Metal spark perforation of the tympanic membrane with deafness and facial paralysis. *J. Laryngol. Otol.*, 699-700 (1986).

(168) Starck, J. and Pekkarinen, J., Industrial impulse noise: crest factor as an additional parameter in exposure measurements. *Appl. Acoustics*, *20*, 263-274 (1987).

(169) Stern, F. B., Waxweiler, R. A., Beaumont, J., Lee, S. T., Rinsky, R. A., Zumwalde, R. D., Halperin, W. E., Bierbaum, P. J., Landrigan, P. J., and Murray, W. E., A case-control study of leukemia at a naval nuclear shipyard. *Am. J. Epid.*, *123* (6), 980-92 (1986).

(170) Stern, R. M., Cancer incidence among welders: possible effects of exposure to extremely low frequency electromagnetic radiation (ELF) and to welding fumes. *Environ. Health Perspect.*, *76*, 221-9 (1987).

(171) Stern, R. M., Drenck, K., Lyngenbo, O., Dirksen, H., and Groth, S., Health, occupational exposure, and thoracic magnetic moment of shipyard workers. In: *Health Hazards and Biological Effects of Welding Fumes and Gases.* Proceedings International Conference, Copenhagen, 18-25 Feb. 1985. Ed: R. M. Stern, A. Berlin, and J. Jarvisalo. Elsevier Science Publishers, Amsterdam, 403-408 (1986).

(172) Stern, R. M., Lipka, J., Madsen, M. B., Morup, S., Thomsen, E., and Koch, C. J. W., Instability of lung retained welding fumes: A slow active Ni and Cr compartment. *Occupational and Environmental Chemical Hazards*, Eds.: V. Foa, E. A. Emmett, M. Maroni, and A. Colombi. Chichester, Ellis Horwood Limited, 444-460 (1987).

(173) Syversen, T. and Jenssen, J., High hydrostatic pressure potentiation of the toxic effects of

chromate in cell culture. *Undersea Biomed. Res., 14* (1), 11-19 (1987).

(174) Tandon, R. K., Ellis J. Crisp, P. T., Baker, R. S., and Chenhall, B. E., Chemical investigation of welding fumes from hardfacing and HSLA-steel electrodes. *Welding J., 65*, 231s-236s (1986).

(175) Thomas, T. L. Stolley, P. D., Stemhagen, A., Fontham, E. T. H., Bleecker, M. L. Stewart, P. A., and Hoover, R. N., Brain tumor mortality risk among men with electrical and electronics jobs: A case-control study. *J. Natl. Cancer Inst., 79* 233-238 (1987).

(176) Thomsen, E., Stern, R. M., and Pedersen, B., Exposure monitoring and chemical analysis of welding fume. In: *Health Hazards and Biological Effects of Welding Fumes and Gases.* Proceedings International Conference, Copenhagen, 18-25 Feb. 1985. Eds: R. M. Stern, A. Berlin, and J. Jarvisalo. Elsevier Science Publishers, Amsterdam, 47-50 (1986).

(177) Tolgyessy, J., Olah, L., and Klehr, E. H., Contribution to the radioanalytical determination of heavy metals in the working environment of welders. *J. Radioanal. Nucl. Chem.*, 114(1), 101-104 (1987).

(178) Turut, P., Isorni, M. C., Sicard, C., and Malthieu, D., Macular photoinjury caused by a welding arc on an eye with an implant. *Bull. Soc. Ophtalmol. Fr., 86*(6-7), 857-859 (1986).

(179) Ulrich, L., Malik, E., and Neumanmova, E., Some effects of exposure to chromium and nickel in the aerosol form. *Prac. Lek., 39*(9), 390-395 (1987).

(180) Uniat, L., Olk, R. J., and Hanish, S. J., Welding arc maculopathy. *Am. J. Ophthalmol.*, 102 (3), 394-395 (1986).

(181) United States EPA, Health assessment document of nickel. U.S. Environmental Protection Agency, Office of Research and Development, Environmental Criteria and Assessment Office, Research Triangle Park, NC 27711 (1986).

(182) Valerio F., Puntoni, R., and Lazzarotto, A., Mutagenic activity of fumes produced by gas-cutting during refitting operations in oil tankers. In: *Health Hazards and Biological Effects of Welding Fumes and Gases.* Proceedings International Conference, Copenhagen, 18-25 Feb. 1985. Eds: R. M. Stern, A. Berlin, and J. Jarvisalo. Elsevier Science Publishers, Amsterdam, 315-318 (1986).

(183) Van der Wal, J. F., Further studies on the exposure of welders to fumes, chromium, nickel and gases in Dutch industries: plasma welding and cutting of stainless steel. *Ann. Occup. Hyg., 30*(2), 153-161 (1986).

(184) Verschoor, M., Herber, R., Van Hemmen, J., Wibowo, A., and Zielhuis, R., Renal function of workers with low-level cadmium exposure. *Scand. J. Work, Environ. Health, 13*, 232-238 (1987).

(185) Vogelmeier, C., Konig, G., Bencze, K., and Fruhmann, G., Pulmonary involvement in zinc fume fever. *Chest, 92*, 946-948 (1987).

(186) Voitkevich, V. G. and Fedorina E. E., Determination of the concentrations of chromium and its compounds in the welding dust from stainless steel welding. *Avtom. Svarka, 6*, 52-55 (1987).

(187) Voitkevich, V. G., and Senkevich, A. I., Study of the heterogeneity of the composition of welding dust particles by x-ray photoelectron spectroscopy. *Avtom. Svarka, 3*, 34-38 (1987).

(188) Wilmer, J. W. G., Reuzel, P. G. J., Hooftman, R. N., and Dreef-Van der Meulen, H. C., The assessment of the carcinogenicity of chromium and nickel containing welding fumes. Phase I: In vitro and in vivo genotoxicity and cytotoxicity studies. *Rapp. - TNO*, Hoofdgroep Voeding Voedingsmiddelen TNO, ISS V 86.362/130370, (1986).

(189) Winburn, D. C., Laser protective eyewear: A critique. *Welding J., 65*, 33-37 (1986).

(190) Wiseman, L. G. and Chapman, E. T., Test program to determine whether nickel carbonyl forms during the welding of nickel-containing alloys. A progress report. In: *Health Hazards and Biological Effects of Welding Fumes and Gases.* Proceedings International Conference, Copenhagen, 18-25 Feb. 1985. Eds: R. M. Stern, A. Berlin, and J. Jarvisalo. Elsevier Science Publishers, Amsterdam, 497-533 (1986).

(191) Wittgens, H., Eyedrops for arc welders. *Arbeitsmed. Sozialmed. Praventivmed.* (Germany, West), *21*(2), 49-50 (1986).

(192) Zober, A., Expert reporting on airway and lung diseases in welders. *Arbeitsmed. Sozialmed. Praventivmed.* (West Germany), *21*(4), 89-92 (1986).

(193) Zschiesche, W., Emmerling, G., Schaller, K. H., and Weltle, D., Investigations on renal impairments of high alloy steel welders. In: *Occupational and Environ. Chemical Hazards*, Eds. V. Foa, E. A. Emmett, M. Maroni, and A. Colombi. Chichester, Ellis Horwood Limited. 339-343 (1987).

(194) Zschiesche, W., Wilhelm, E., Zober, A., Schaller, K. H., Weltle, D., and Valentin, H., Manganese in stainless steel welding fumes: external exposure and biological monitoring. In: *Health Hazards and Biological Effects of Welding Fumes and Gases.* Proceedings International Conference, Copenhagen, 18-25 Feb. 1985. Eds: R. M. Stern, A. Berlin, and J. Jarvisalo. Elsevier Science Publishers, Amsterdam, 193-196 (1986).